No Sea Too Deep

The History of Oceanographic Instruments

Anita McConnell

Curator of the Oceanography Collection
Science Museum, London

Adam Hilger Ltd, Bristol

British Library Cataloguing in Publication Data

McConnell, Anita
No sea too deep: the history of oceanographic instruments.
1. Oceanographic instruments—History
I. Title
551.46 GC41

ISBN 0-85274-416-1

Consultant Editor: **Professor A J Meadows,** University of Leicester

Published by Adam Hilger Ltd, Techno House, Redcliffe Way, Bristol BS1 6NX.
The Adam Hilger book-publishing imprint is owned by The Institute of Physics.

Typeset by Quadraset Ltd, Radstock, and printed in Great Britain by J W Arrowsmith Ltd, Bristol.

To Connie

Who never fails to keep me afloat
in life's storms

Preface

The history of marine science has been related by several of the men who were themselves responsible for part of its progress. For example, Wyville Thomson (1873) gave an account of the physical and zoological researches undertaken prior to his cruises in HM Ships *Lightning* and *Porcupine* during the summers of 1868 to 1870. John Murray prefaced his narrative of the *Challenger* circumnavigation in 1872–76 with an account of the scientific advances that had preceded it (*Report Narrative* **1** xxxi–xlix); he and Johan Hjørt (Murray and Hjørt 1912) covered much the same ground when relating the story of the researches made on board the *Michael Sars* in 1910. G H Fowler (1912) also gave a brief historical introduction to his textbook *The Science of the Sea*.

More recently, a revival of interest has led to three conferences on the history of oceanography, the first of which was held in Monaco in 1966 (*Bull. Inst. Oceanogr.* 1968), the second in Edinburgh in 1972 (*Proc. R. Soc. Edin.* 1972) and the third in Woods Hole, MA, in 1980 (Sears and Merriman 1980). The most detailed study of the subject is that of Margaret Deacon (1971), who covered all aspects of marine science from 1650 to 1900. The American viewpoint has been treated by Susan Schlee (1973) and Vincent Ponko (1974), while the history of the International Commission for the Exploration of the Sea (ICES)—from its foundation in 1902 until the present time—has been chronicled by Arthur Went (1972), one of its former Presidents.

All these writers have discussed the aims of the scientists concerned and have shown how their conceptions of conditions within the seas and oceans were gradually made more realistic as imagination was replaced by data of ever-increasing reliability. It is not my intention to go over the same ground but rather, by looking at original publications, to discover how this knowledge of conditions within the deep sea was obtained. I have tried to locate the technical descriptions of apparatus used for measuring the depth of the sea and for recovering samples of the sea bed; and those for collecting samples of sea water from known depths. I have also attempted to trace the development of the deep-sea thermometer and meters for measuring submarine currents. I have been concerned to note how the apparatus was actually used, whether it fulfilled the hopes of its designers, and if it led to improved forms. I have also considered the inputs from technical advances that were being made simultaneously in unrelated fields of engineering.

The major portion of this book is concerned with the period beginning in 1800 when a variety of apparatus was constructed and tested in deep-water conditions. However, many contemporary descriptions referred to much earlier work done during the late seventeenth and early eighteenth centuries. It is therefore necessary to introduce the ideas of Boyle, Hooke and Hales, and to determine the extent to which their ideas were transmitted through the relatively unproductive eighteenth century to inspire the designers of those instruments sent out with the Arctic voyages of the early nineteenth century.

The richness of available material has compelled me to be selective. I have chosen to illustrate the general apparatus of the early nineteenth century with reference to the Arctic expeditions sent out from Britain. The development of sounding techniques is linked to the work of the telegraph companies who were responsible for many of the advances made in this field. The waterbottle and the thermometer were the prime tools of marine zoologists, and their evolution is related to this branch of the science in particular. Scandinavian oceanographers pioneered the modern approach of detailed synoptic mapping, devising for this purpose a variety of precision waterbottles. The current meter, the last arrival on the scene, is treated in more general terms.

Contemporary descriptions, initially published in journals of the learned societies, transferred in later centuries to the newly founded technical magazines. A rising public interest was met by illustrated descriptions in books and encyclopedias. At first, having no obvious commercial application, oceanographic instruments were not protected by patent. When apparatus was eventually designed and made in some quantity it was often patented and was advertised at exhibitions and in manufacturers' catalogues.

Such a range of potential reference material, reflecting the interests of many nations at one period or another in some of oceanography's uses, can hardly be scanned by one individual. I hope that this study will encourage others to uncover more details of man's ingenuity, by which we may now justifiably claim that there is 'No Sea Too Deep'.

Anita McConnell
Science Museum, London

REFERENCES

Wyville Thomson C 1873 *The Depths of the Sea*

Murray J 1880 *Challenger Report Narrative* 1880

Murray J and Hjørt J 1912 *The Depths of the Ocean*

Fowler G 1912 *Science of the Sea*

Premier Congrès International d'Histoire de l'Océanographie, Monaco 1966 *Bull. Inst. Oceanogr.* 1968 (2 vols).

Second International Congress on the History of Oceanography, Edinburgh 1972 *Proc. R. Soc. Edin* 1972 **72** and **73**

Sears M and Merriman D (Eds) 1980 *Oceanography: the past* (New York: Springer-Verlag)

Deacon M 1971 *Scientists and the Sea 1650–1900*

Deacon M 1971 *Scientists and the Sea 1650*–1900

Schlee S 1973 *A History of Oceanography*

Ponko V 1974 *Ships, Seas and Scientists*

Went A 1972 'Seventy years a-growing: a history of ICES' *Cons. Perm. Int. Expl. Mer. Rapports et Procès-verbaux* **165** 232 pp

Acknowledgments

Apart from the sources listed in my references I owe a great deal to colleagues in Britain and abroad who showed me the historical instruments in their collections, who searched their library shelves to locate the manuscripts and books that I required, who helped me with translations, and who answered many technical enquiries in fields beyond my immediate knowledge. Foremost among these I wish to thank particularly Margaret Deacon, Arthur Mills, Joe Roome and Ian Carter.

Others, too numerous to mention individually, are the directors, curatorial staff and librarians of the following institutions:

The Science Museum, London
The National Maritime Museum, Greenwich
The Oceanographic Museum, Monaco
The Smithsonian Museum of the History of Technology, Washington
Woods Hole Oceanographic Institute, Woods Hole, Massachussets
The Royal Scottish Museum, Edinburgh
The Geophysical Institute, Bergen
The British Library, London
The Royal Society, London
The Royal Geographical Society, London
The Admiralty Historical Library, London
Messrs Cable and Wireless, London
Whitby Museum, Whitby, Yorkshire
The Naval Museum, Rio de Janeiro

Acknowledgments are due to the following for their permission to reproduce illustrations:

Figures 3, 5: Reproduced by permission of the Royal Society, London.
Figures 6, 8, 21, 37, 39, 41, 44, 46, 49, 52, 55, 74, 82, 84, 91, 94, 97, 99, 101, 105, 114, 116, 123(*a*), 112, 129, 140, 149, 152: Crown Copyright, Science Museum, London.
Figure 88: Reproduced by permission of the National Maritime Museum, Greenwich.

Considerable practical help towards the production of this book came from the Science Museum workshops, photostudio and design office, while the efficient cooperation of Neville Goodman and Valerie Jones of Adam Hilger Ltd is also much appreciated.

To all these, and to all who over the last few years have offered counsel, advice and encouragement, my thanks.

Contents

Chapter 1

Man and the Ocean

The oceans cover 70% of our planet and the volume of water is far greater than that of the land above sea level. Exploration of the entire sea is therefore an enormous undertaking both in terms of its area and the depth of water from surface to sea bed. The problems involved in measuring and recording the mass of information needed to describe fully this marine realm did not become apparent all at once to the early scientists. Rather, as the frontiers of exploration were pushed out from the shore and down into the abyss, new questions arose and new techniques had to be devised if answers were to be found.

It seems useful to summarise what we know today about the oceans, while bearing in mind that less is known about the sea bed than about the surface of the Moon.

We can speak of a 'world ocean' in the sense that it is possible to sail continuously around the world, but usually we define three major oceans—the Pacific, Atlantic and Indian Oceans—which are largely separated by continents so that circulation of their waters is restricted. Oceanographers find the conventional geographic divisions of three major and many minor oceans and seas useful because the nature of the water differs within each of these areas.

The principal topographic divisions of the oceans are the *continental margins,* the *abyssal plains,* and the *oceanic ridges*. Continental margins, comprising shelves, slopes and rises, are submerged extensions of the land masses. The continental shelf extends on average 50 km from the shoreline, although this distance can vary considerably, and is covered with sediments derived largely from nearby land, together with detritus from marine organisms. The shelf terminates abruptly at the continental slope, which drops fairly steeply to considerable depths of the order of 2000–3000 m. From here the continental slope levels out gently, meeting the abyssal plains at depths of 4000–5000 m; these plains make up over one-third of the area of the Atlantic and Indian Ocean basins and three-quarters of the Pacific basin. Blanketed by fine clays, they are featureless apart from volcanic islands and submerged 'sea mounts' which often occur in clusters connected on the sea floor by the accumulations from volcanic activity and erosion. The oceanic ridge systems form a series of interconnected ranges rising 2000–4000 m above the ocean floor; they represent the margins of the tectonic plates which make up the Earth's crust. In some parts of the world molten core material rises to the surface, spreads out and solidifies to create new areas of ocean floor. This generation of new floor is counterbalanced elsewhere by movement of the crust down into the mantle. The steep-sided trenches created where the sea floor is pulled down are the deepest parts of the ocean; the distance from the sea surface to the floor of a trench may be over 10 000 m. The topographical range within the oceans is therefore far greater than that on land where the highest peak, Mount Everest, rises a mere 8884 m above sea level.

Sea water, with its characteristic ratio of dissolved salts, is believed to have been accumulating throughout geological time. New brines accompany basaltic rocks pushed up at ocean ridges, and are also derived from volcanoes. The salinity of sea water is measured as the total number of grams of dissolved salt in one kilogram of sea water (parts per thousand), and in the open sea this can range from 33 to 38‰. Where rivers discharge into a restricted arm of the sea, such as in the Baltic Sea, the water is considerably less saline than that of the open ocean. The major constituents, which are mostly present in ionic form, are sodium, potassium, calcium and magnesium, although almost all elements are present in sea water in very small amounts. There are also variable quantities of nutrients such as nitrogen compounds, phosphate, silicate and oxygen in its gaseous form. These, together with water and bicarbonates, make up compounds essential to all plant and animal life.

The world ocean functions as a gigantic heat pump, transferring heat from the equator to the poles. Surface water flow is guided by the winds and by forces due to the Earth's rotation and the shape of the ocean basins; for example, in the middle latitudes of both hemispheres the prevailing westerly winds set up an eastward drift. In the southern hemisphere this drift is not interrupted by land masses and so circulates round the continent of Antarctica as a continuous westward drift. In the northern hemisphere the continents obstruct the drift, deflecting it both north and south. Thus a vast clockwise eddy, of which the Gulf Stream forms an important part, is established in the North Atlantic. The South Atlantic and other oceans have similar eddies or 'gyres'. Many of the strong surface currents are underlain in mid-water by counter-currents flowing in the opposite direction to those above.

When the surface currents reach polar latitudes they are chilled and their salinity is increased by the formation of ice from which most of the salt is excluded. The remaining water thus becomes very dense and sinks to the bottom, returning towards the equator as a slow drift through the deeper parts of the ocean.

The global circulation of heat results in a temperature profile through the water with three main divisions. There is a surface or mixed layer, a few tens or hundreds of metres thick, which reflects the average air temperatures at that latitude; below this is an intermediate 'thermocline' region where there is a fairly abrupt reduction in temperature with depth; and thirdly there is a deep bottom layer, of polar origin, with temperatures only a few degrees above the freezing point of sea water (around −2 °C, depending on salinity). This deep bottom water is normally found at depths below 1000–2000 m.

If two distinct masses of sea water are brought together by opposing currents, then the more saline, denser one will lie beneath the less dense one. If the masses are of identical salinity, then the colder water will be more dense; conversely, if both masses are at the same temperature, then the more saline water will be more dense. In polar seas, therefore, the surface water is often colder than that at depth, since it is usually less saline due to dilution by ice meltwater and river run-off from adjacent land masses. In temperate and tropical seas the water temperature usually decreases from surface to sea bed, although in all seas tongues of water contrasting sharply in temperature or salinity may occur as permanent or ephemeral features.

Within the sea, pressure increases with depth by one atmosphere every 10 m, so that at a depth of 5000 m, on the abyssal plains, the pressure is 500 times greater than at the surface. In the deepest oceanic trenches, 10 000 m down, the pressure is over 100 kg/mm^2.

Plant life flourishes wherever sunlight penetrates the water, forming the primary stage in the marine food chain. As on land, both fertile regions and deserts are found, depending on the presence or absence of nutrients. Generally speaking, shallow seas are rich feeding grounds, as are those zones where water wells up from depth, carrying nutrients with it. Fishing grounds of major economic importance to man are therefore to be found over the continental shelves, and the yields or failures of such fisheries are of great concern to the nations which have traditionally exploited them. Elsewhere life may be more sparse, yet although much of the ocean's volume is pitch dark, cold and under great pressure, life has adapted to this environment and marine organisms are to be found in every region, both in mid-water and on the sea floor.

The debates of the early philosophers, prior to the seventeenth century, as to conditions within the sea could perhaps be more properly classed as religious speculation. Their approach to the matter was conditioned, not by science as we now understand it, but by their desire to prove that the world was the result of some deliberate act of creation by a rational deity, and not just a chance occurrence. The oceans reflected or counterbalanced the land and, though they might shelter creatures (real or mythical), they were at best a source of livelihood for a few sailors or fishermen, and at worst an elemental force to be feared and avoided.

A more scientific approach to the study of the world's oceans was begun in seventeenth-century Europe, but it was still impossible for the early marine scientists to discover the boundaries of the oceans and the ranges of temperature and salinity within their waters. For, when this interest began to develop, the geographical exploration of the surface of the ocean was still incomplete, and speculation on conditions at the poles and in the great southern ocean ran unchecked until the close of the eighteenth century. By then cartographers had been able to draw in most of the coastlines and all that remained was to probe into those regions to which no sensible merchantman or armed fighting ship would normally venture.

As one would expect, the degree of accuracy demanded of oceanographic apparatus increased as the years went by. The improvement was not linear, but moved in bursts as first one parameter and then another became crucial for some aspect of marine science or technology. To begin with, very crude measurement sufficed. To discover if the sea is 'fathomless' or bounded is not a matter of getting a sounding accurate to a few metres but of the technical ability to get a line or apparatus from surface to sea bed, and back. Time enough, when this can be accomplished, to consider whether a truly vertical distance has been measured. Similarly, chemists pondered just how salt the sea might be in tropical regions where, they wondered, to what extent the sun's heat might evaporate the surface layer of water to leave a very salty residue. The measurement of salinity (it was soon discovered that it was much the same everywhere in the open sea) was sufficiently accurate to begin with. The same held true of temperature measurement within the sea. The great depths were not frozen solid, nor were they heated by their proximity to the centre of the Earth, as the rocks below volcanoes apparently were.

In this way simple boundary concepts of volume, temperature and salinity began to take shape, and when the nineteenth century dawned it became possible to improve instruments such as sounders, thermometers and waterbottles because there was a broad idea of the range of conditions over which they might be called on to register. However, the frontiers of exploration still lay relatively close to the water surface, and the difficulties of operating apparatus under great pressure and at considerable distances below the mother-ship were not easily overcome. Most instruments therefore failed to register at the hoped-for precision.

Necessity is the mother of invention and, in step with technical progress in other fields throughout the nineteenth century, one instrument after another was improved. But, though it was common practice to sound, to take water samples, and to measure temperature at each observation point, the same improvements were not looked for in all three classes of instrument. The layout of this book therefore reflects this lopsided progress. Thus for submarine telegraph cable laying, precise depth measurements and the collection of bottom samples were essential, while temperature and salinity measurements were less important; average values sufficed for these purposes. In complete contrast, for marine zoology to progress, waterbottles needed improvement and thermometers had to be accurate to fractions of a degree, a standard they notably failed to achieve throughout most of the nineteenth century. During the early years of the twentieth century, work was being undertaken in the sea areas of greatest complexity, in the North Atlantic and adjacent polar waters, and for this thermometers sensitive to hundredths of a degree were needed and were indeed made.

The major features of the sea floor, the ridges and troughs, and the smaller topographic landforms such as sand waves, were not discovered until a large number

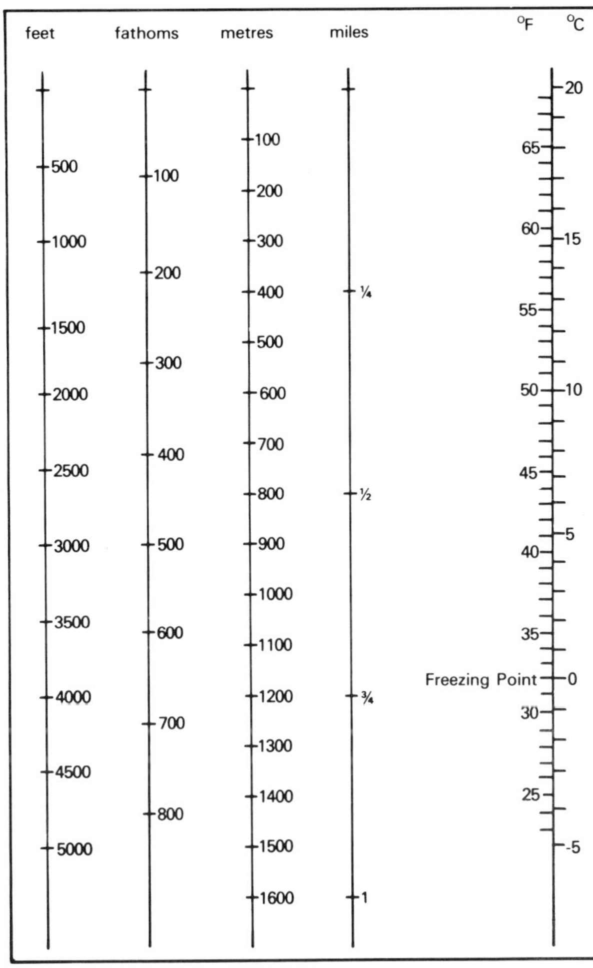

Figure 1. Conversion table for depths and temperatures. Shelf seas are less than 180 m; mean ocean depths are 600 m; the greatest depth known at present is 13 000 m. From British Standard 350 Pt 2 (1962).

of soundings had been made and the accumulated data assembled on charts. If the spacings between soundings are greater than the area that such features occupy, as in the case of ridges and troughs which are long and narrow, then the survey will fail to register their presence. Their typical form and size will therefore not be appreciated and in fact the pattern of ridges and troughs was not understood until the turn of the twentieth century, while the smaller sand waves had to await the development of the continuous echo-sounding profile before they could be placed on the charts.

Chapter 2

The Pre-technological Era, 1660–1770

2.1. ROBERT BOYLE: OCEANOGRAPHY WITHOUT INSTRUMENTS

The Honourable Robert Boyle (1627–91) was one of the great seventeenth-century scientists. His interests ranged widely: he maintained a laboratory for physical and chemical experiments and he wrote many books, including a small volume entitled *Tracts*, which appeared in 1671 (figure 2).[1] One of its five sections was 'The Temperature of the Submarine Regions', and another was 'The Bottom of the Sea'. Boyle, though strictly a landsman, had not been content to remain in his laboratory or study and draw conclusions unsupported by evidence. He sought and obtained facts by questioning seamen and divers about temperatures below the surface, adding to these some information obtained from experiments of his own. His Introduction revealed the prevailing ignorance:

> . . . what I shall say will probably spoil the Vulgar Error in that all deepe water of which the Sea is the Chiefest, the lowermost are still the warmest parts, unlesse . . . the superficiall ones happen to be a little warm'd by the Extraordinary or Violent heat of the Sun.[2]

His evidence was based on the account of a transatlantic crossing:

> . . . the following Passage, taken out of the Voyage of Monsieu de Monts into New-France . . . About the eighteenth day of June we found the Seawater during three days space very warm, and by the same warmth, our wine was also warme in the bottom of our ship; yet the Air was no hotter than before. And the 21 of the said month, quite contrary, we were 2 or 3 days so much compassed with mists and cold, that we thought ourselves to be in the month of January, and the water of the sea was extremely cold: which continued with us untill we came upon the Bank by reason of the said mists which outwardly did procure this cold unto us.[3]

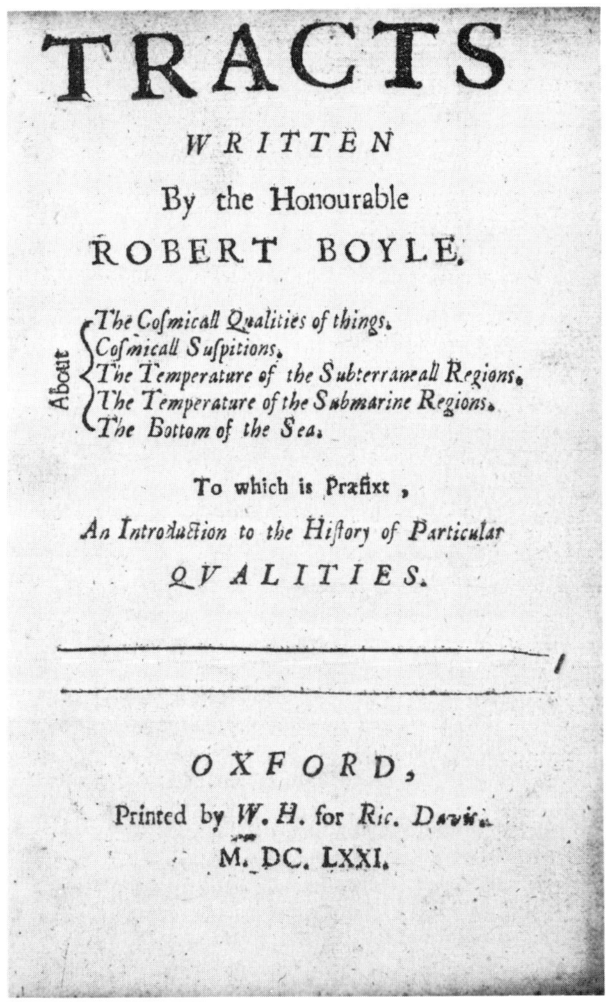
TRACTS

WRITTEN

By the Honourable

ROBERT BOYLE.

About
The Cosmicall Qualities of things.
Cosmicall Suspitions.
The Temperature of the Subterraneall Regions.
The Temperature of the Submarine Regions.
The Bottom of the Sea.

To which is Præfixt,

An Introduction to the History of Particular QVALITIES.

OXFORD,

Printed by W. H. for Ric. Davis.

M. DC. LXXI.

Figure 2. Title page of Boyle's *Tracts*.

Having dealt with the upper layers,

> . . . it remains that I take notice of the Temperature of the Lower Region, which, in one word, is Cold (unlesse in some few places to be presently mentioned).[4]

His information on deep water temperatures came

> from persons that have occasion to goe down into it, or otherwise take notice of its Temperature in very differing Regions of the World, and at very unequall depths.[5]

Questioning a diver salvaging merchandise from sunken vessels in northern waters, he learned that it was the coldness that limited the time that they could work below the surface. Other correspondents told him that in American and tropical African waters they found that the cold increased with depth.

A diver's range was limited, but Boyle was able to extend his observations

> [On a vessel sailing to the East Indies] whilst they pass through the Torrid Zone, to keep their Drink, whether wine or water, cool, by letting it down in bottles to the depth of 80, 90 and sometimes an hundred Fathom or better, and letting it stay there a competent time; after which he told me he found it to be exceeding cool and refreshing.[6]

And deeper still:

> Lastly, to satisfy myself as far as I could, to how great a depth the Coldnesse of the Sea reached; meeting an observing Traveller whose affairs or Curiosity has carried him to divers parts both of the East and West-Indies, I enquired of him whether he had taken notice of any extraordinary deep soundings in the vaster seas, to which being answered, that some years agoe, sailing to the East Indies . . . they had the Curiousity to let down 400 fathom of Line and found that they needed no lesse. Whereupon I enquired of him, whether he had taken notice of the Temperature of the Sounding Lead as soon as 'twas drawn up: To which he told me, that he, and some others did; and that the Lead which was of the weight of about 25 or 30 lbs, had received so intense a degree of coldnesse as was very remarkable; insomuch that he thought that if it had been a masse of Ice, it could not have more vehemently refrigerated his hands.[7]

Boyle's information, gathered without the use of a thermometer, was of necessity only relative. Yet such careful questioning gave him sufficient evidence to draw the correct conclusion that

> Though the Air and the Earth have been discriminated as to Temperature into three Regions, yet the Informations I have hitherto met with, invite me not to assign to the Sea any more than two. The former of which may be supposed to reach from the superficies of it, as far downward as the manifest operation of the variously reflected and refracted Beames of the Sun, or other Causes of warmth penetrates; from which to the Bottom of the Sea, the other region may be supposed to extend.[8]

He then suggested that a 'Weatherglass' might be adapted to test submarine temperatures but having discovered by trial that it responded more to pressure than to temperature, concluded that it was unfit for that purpose, a caution that some of his successors could have profitably borne in mind.

The following tract, entitled 'The Bottom of the Sea', used the same method of enquiry. Mariners confirmed Boyle's supposition that in shallow water the sea bed was by no means a smooth bowl, but a landscape with its own topography.

2.2. ROBERT HOOKE: THE SCIENTIST GOES TO SEA

Robert Hooke (1635–1703) had been Boyle's laboratory assistant before coming to the Royal Society in 1662 as its first Curator of Experiments. It was his task to design and construct apparatus to test hypotheses of interest to members. The first known drawings of a sounder and a sampler to collect sea water from beneath the surface are preserved in the Society's collection of Hooke's papers, dated September 1663 (figure 3).[9]

It seems that certain members of the Society had for some years been experimenting with alternative methods of sounding, even going to the Thames estuary to take part in trials of various apparatus. The first published indication of their interest in the practicalities of marine science appeared in the *Philosophical Transactions* in January 1666, which took the form of *Directions for Seamen Bound for Far Voyages*.[10] Compiled some years previously by Laurence Rooke who prior to his death in 1662 had been Professor of Astronomy at Gresham College where the Society's rooms were also situated, the *Directions* listed compass variations, tidal, charting and meteorological observations as particularly important. These were to be logged and submitted on return to the Lord High Admiral, the Duke of York and Trinity House, the latter passing the information to the Royal Society. Of the sea itself they were required

> . . . to take notice of the Ground at the bottom of the Sea, in all soundings, whether it be Clay, Sand or Rock etc. To carry with them good Scales and Glass-violls, of a pint or so, with very narrow mouths, which are to be filled with sea-water in different degrees of latitude, as often as they please, and the weight of the Vial full of water taken exactly at every time, marking withal the degree of latitude and the day of the month: and that of the water near the Top, as at greater Depth.

These specific gravity measurements were needed to discover if the sea was saltier at the bottom and if, as some mariners claimed, freshwater springs emerged at the sea floor.

The following issue of *Transactions* carried an *Appendix* to the *Directions.*[11] Detailed instructions were given for constructing the necessary apparatus, together with an illustration based on Hooke's earlier sketch (figure 4), so that any ship's carpenter and smith could have made the instruments as specified. The sounder consisted of a wooden float with a wire spring on which hung a sinker of stone or iron. Dropped over-

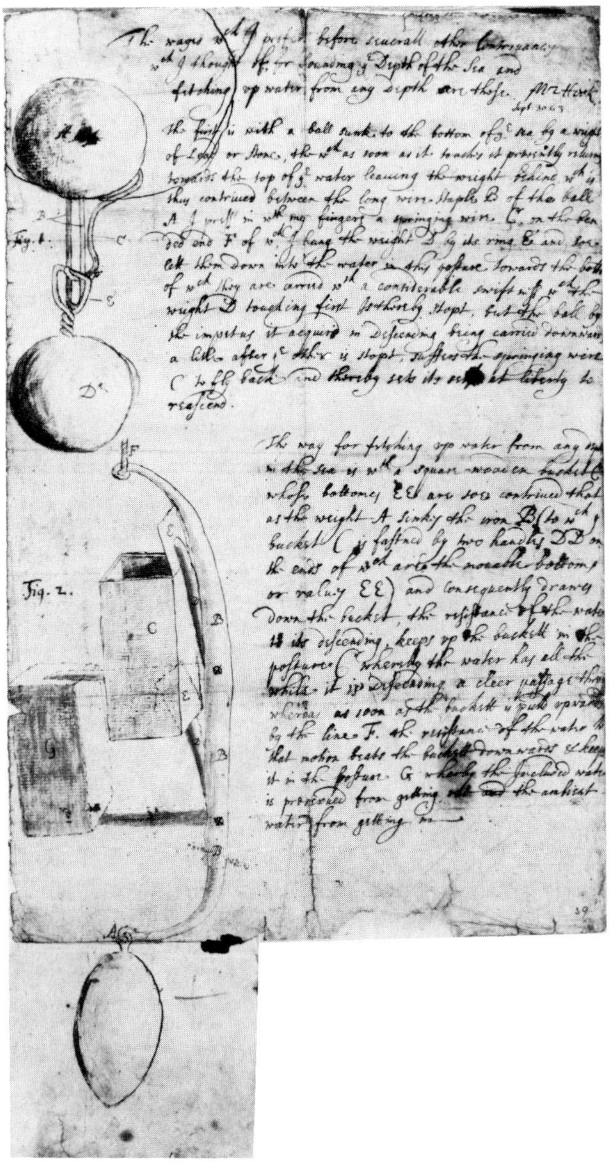

Figure 3. Hooke's detaching-weight sounder and water sampler, sketched by himself. From Hooke Mss XX No. 23, 1663.

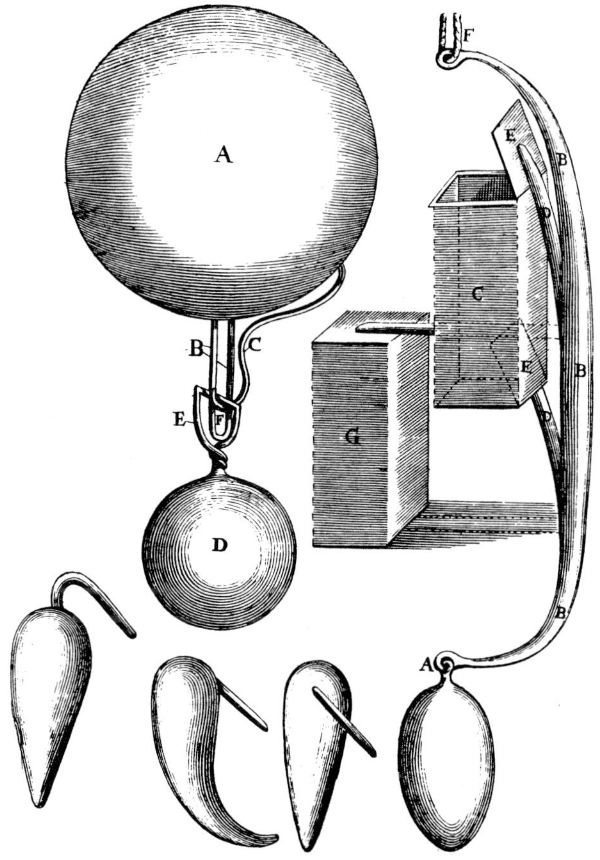

Figure 4. Hooke's detaching-weight sounder and water bucket. Below are shown three alternative forms of detaching weight for the sounder. From *Phil. Trans.* 1(24), 1667.

board, this assembly sank to the sea bed. There the wire being relieved of the sinker's pull, it sprung apart, allowing the sinker to fall off. The float then rose to the surface. By timing its absence with a pendulum and comparing this to its performance in known depths, it was considered possible to sound beyond the reach of lead and line. When trials were carried out in 19 feet of water, six seconds elapsed between the release and reappearance of the sounder, a time interval which remained constant even when the float was swept away by the tide-stream and surfaced some distance from where it had been released. The instructions emphasised the care needed to balance the respective pulls of sinker and float since the depth was calculated from half the travel time, and it was essential that the apparatus should travel down and up at the same speed.

The concept of a lineless sounder on these principles had previously been mentioned by several writers[12] and was certainly no novelty to the well read scientist of Hooke's day. The want of a suitable timing device and the detrimental effect of pressure on any buoyant material would however have prevented such a sounder from working in any but the shallowest water. The sampler described in the instructions was simply a square wooden box attached to a weighted frame. Its flap lids were held open by water flow during its descent. On hauling up the line, the entire box dropped on hinges to a lower position and the lids closed over the contained water.

The publication of these instructions in the *Transactions* was less esoteric than might at first sight appear. Apart from their links with Gresham College and its training of mariners, the Society enjoyed the King's patronage. It lay within his power to command the testing of their apparatus by ships of the Royal Navy.

Unluckily for Hooke, however, the engraving of the sounder reproduced in the *Transactions* was wrongly drawn, and a disclaimer had to be made in a later note to the effect that the detaching spring had been depicted outside rather than inside the float's suspension hook.[13] This led to certain misunderstandings when, in May 1666, an abridged translation was published by the French *Journal des Sçavans*[14] accompanied by a copy of the plate complete with errors of drawing on both sounder and sampler. This in turn drew comment from an engineer, Pierre Petit.[15] He criticised Hooke's description of both instruments, finding both text and illustrations hard to comprehend. It had been necessary to guess that the sampler arms were hinged in order that they might move and be sprung in some way. The French had developed a simpler apparatus which they preferred, and Petit himself had made one of brass with double valves. The sounder was no novelty to him either and he referred to that described by Cusanus in the philosophical treatise *De Staticis Experimentis* of 1450. Further, the true sinking rate of such sounders had not been established as travel times in shallow water could not justifiably be extrapolated to greater depths.

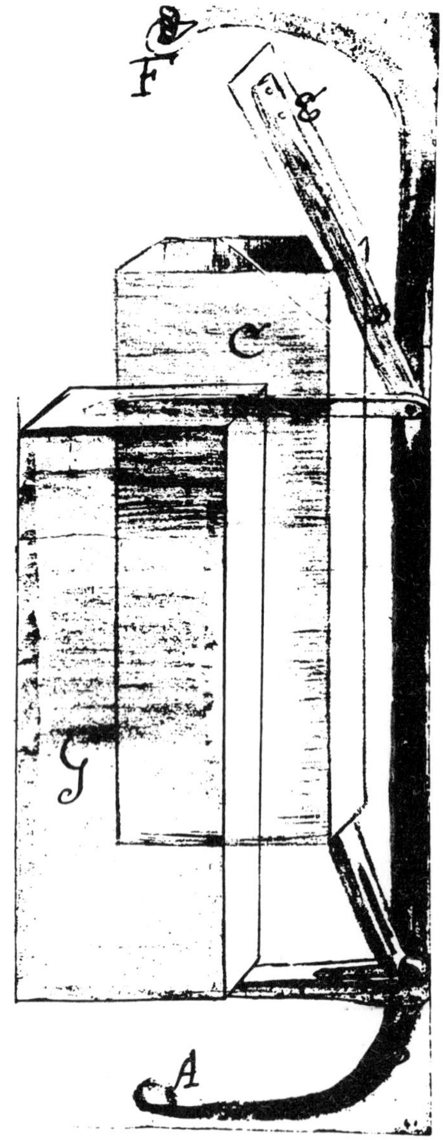

Figure 5. Hooke's sampler, sketched by himself. From Hooke Mss XX No. 35.

Hooke's original sketch for the sampler (figure 2) had shown hinges on the frame but these had been omitted by the engraver, as is evident from figure 4. Hooke later prepared another sketch (figure 5) in which he took pains to show the hinges more clearly.[16] Figure 6 shows a modern replica.

The criticism by the French was taken badly, however, and Henry Oldenburg, Secretary of the Royal Society and Editor of the *Transactions,* was moved to publish an irate reaction.[17] Their lack of comprehension, he wrote, was due to their poor grasp of English; the correction to the sampler engraving had been published in plenty of time, and with regard to the French waterbottle, this had been tried in England and found to be unsatisfactory for its valves did not open fully and so impeded its passage through the water.[18] Moreover, Hooke had never claimed the sounder as his invention, only that the design was new. The sinking rate had been confirmed by the simultaneous use of the conventional sounding line.

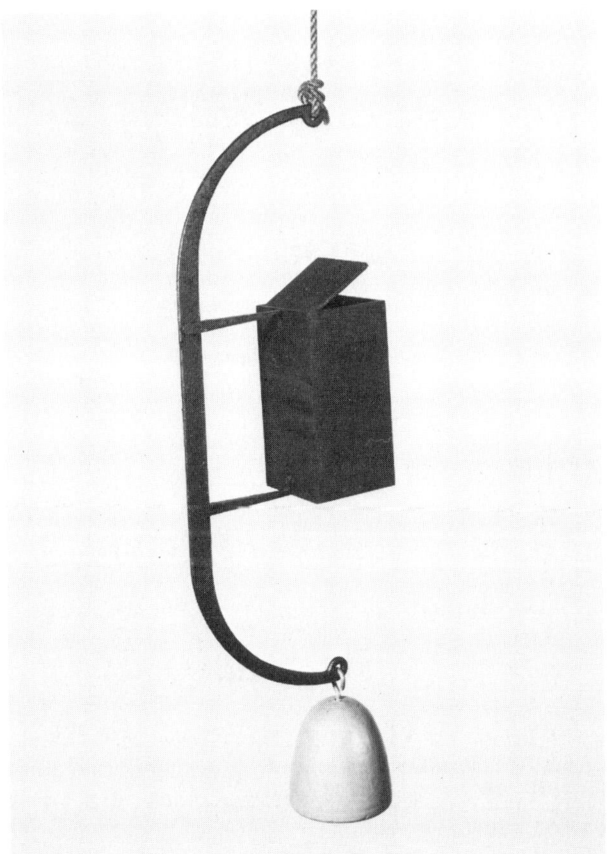

Figure 6. Replica of Hooke's sampler.

In 1667 the instructions were reprinted in the *Transactions* in expanded form;[19] the sounder engraving was corrected but the sampler was still without clearly defined hinges. Possibly Hooke himself had not finally decided how it might best be constructed, for as shown, even with hinges, it is mechanically unsound. It was stated that Mr Richard Shortgrave, Operator to the Royal Society, could supply instruments and advise on their use.

Hooke introduced several new instruments in a course of lectures given in 1691, naming these in florid style 'Emissarius', 'Explorator' and 'Nuntio'.[20] He admitted the difficulty of constructing his original sounder so that it moved downward and upward through the water at the same velocity. Nor could the float be easily sighted the instant it reappeared at the

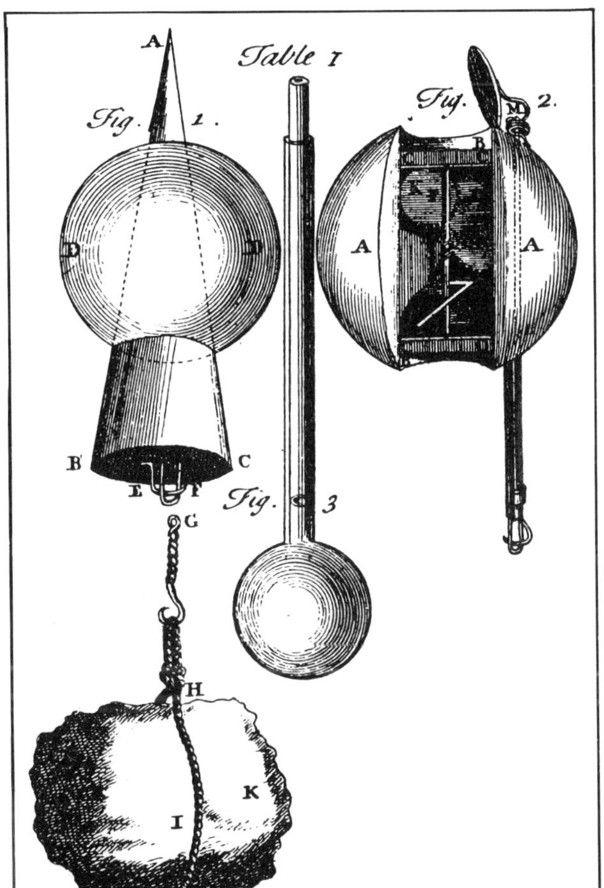

Figure 7. Hooke's pressure sounder, thermoscope and waywiser sounder. From Derham 1726 *Philosophical Experiments of . . . Dr Hooke* table I.

surface (since it was less than 16 inches in diameter and painted black with pitch this is hardly surprising, for seldom are waves less high than this). His new instruments were based on different principles: in the first (figure 7) the pressure of sea water forced entry against air contained in a hollow cone fixed within the float. On recovery, the water drained from the float was weighed; this weight was proportional to the maximum pressure, and therefore the depth, encountered. Thus neither the weight of the sinker nor any delay in sighting the float affected the result.

The mechanism in Hooke's other sounder resembled a land surveyor's waywiser, defined as a vane or ship's log fitted within the float. A vane was turned by water flow, and the number of its revolutions were counted on a register during the descent. On its ascent, a lid automatically dropped over the mechanism so that it became inoperative. Figure 8 shows a modern replica. A thermoscope, intended to show the relative fall in temperature, was also described and illustrated,

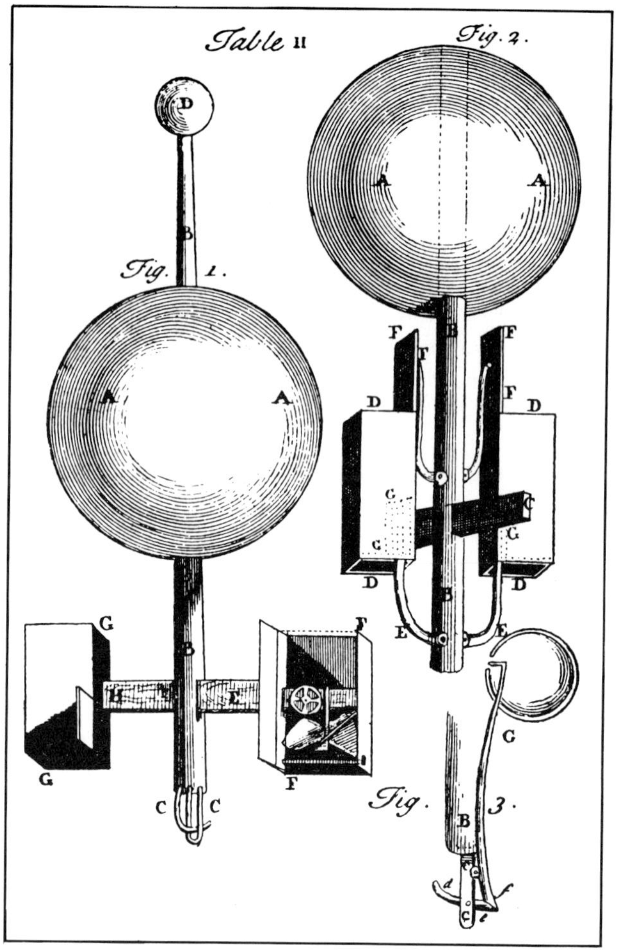

Figure 9. Hooke's detaching-weight sounder adapted to carry two waywiser sounders or two water samplers. From Derham 1725 *Philosophical Experiments of . . . Dr Hooke* table II.

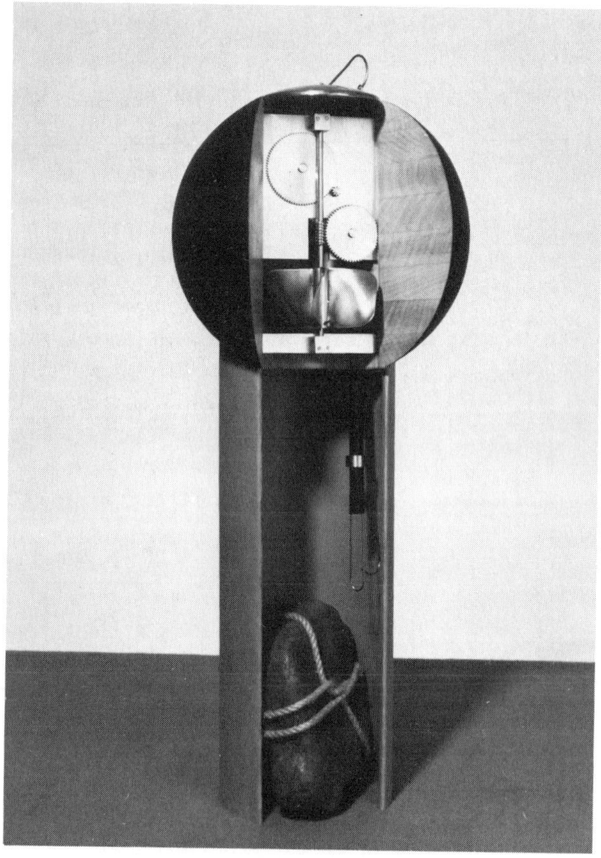

Figure 8. Replica of Hooke's waywiser sounder.

although this would have been susceptible to pressure and vulnerable to wave action, and seems unlikely to have been practical.

The sampler remained unaltered but, together with the mechanical sounder, was incorporated into an apparatus of amazing complexity. To the original float and detaching sinker Hooke now added two sounders, one to work in each direction of travel (figure 9). Two samplers, one to collect water from the bottom, and the other to close on the way up, were carried on cross-arms. It is unlikely that such a device could ever have worked; the drawings are certainly unrealistic, while its operating depth would have been limited by the use of wood in its construction. Hooke had also made clocks and meteorological instruments and seems to have transferred his ideas to marine instruments without taking account of the vastly different environment in which the latter would have to operate. His seafaring was restricted to boat trips in rivers or estuaries, usually made in fine weather.

The value of Hooke's work lies less in its originality, which he himself disclaimed, than in the detailed descriptions which became widely known through the medium of the *Philosophical Transactions.* Over the next two centuries the original engravings were reproduced many times in books and encyclopedias, while instruments owing much to his designs were also produced, such as the pressure sounder developed by Nicholson a century later, for example.[21]

Hooke set out the three principles by which depth could be sounded: by timing a lineless detaching-weight apparatus; by measuring pressure as equivalent to depth; and by using a vane geared to a counter to register the distance travelled. All these principles served, with greater or lesser degrees of success, until the last quarter of the nineteenth century when it became technically feasible to measure depth directly with a wire line.

By the close of the seventeenth century enthusiasm for investigating the natural world had waned in Britain, as it had done in France some years previously. The Royal Society fell on hard times and the apparatus kept in its rooms at Gresham College lay in a state of neglect, so that any of Hooke's surviving instruments were probably lost at this time. When its fortunes revived during the eighteenth century the Society turned its attention to such new topics as progressive agriculture and industry. Such useful and financially rewarding arts of manufacture were seen as more appropriate for study than abstract matters of the Universe which had in any case been comprehensively treated by Isaac Newton. While new interests such as electricity and magnetism were irrelevant to marine science, others such as chemistry could and did have offshoots: there were profitable investigations into the nature of sea water and its variations in different regions of the ocean. Physical experiments such as those of Stephen Hales (see §2.4) were relevant when considering the possible effects of pressure at great depth. The experiments of Marsigli, which are described below, may be seen as the last great feats of the early scientific revolution, and British marine science made only limited progress until in 1773 the Navy took an interest when the Admiralty despatched an expedition to the Arctic.

2.3. LUIGI-FERDINANDO DI MARSIGLI: HYDROGRAPHIC STUDIES

Count Luigi-Ferdinando di Marsigli (1658–1730)[22] was born in Bologna, Italy, and attended its university to study mathematics and natural history. In 1679 he was sent to Constantinople on diplomatic service, and put his knowledge to use by embarking on a study of the Bosphorus Strait.[23] He was particularly intrigued by its system of currents, hearing from local fishermen that the surface flow from the Black Sea into the Sea of Marmara overlay one moving in the reverse direction. He confirmed these reports by the use of a simple apparatus. The surface current was gauged from the Tower of Leander which stood a little way out from the Asian shore. There he set a wooden machine having a six-bladed wheel held in a frame, and counted the number of its revolutions against 100 swings of a pendulum. He counted 38 revolutions, although after four turns one of the blades broke loose and impeded the flow, and a light wind arose to disturb the water surface. He estimated that the stream velocity within the open channel was one-third greater than that at the Tower.

Fishermen were made aware of the submarine current by its drag on their nets, and Marsigli detected it by similar means. From a boat he first lowered a string of white-painted corks (figure 10) which could be seen to swing in contrary directions within each current. Next he lowered a weighted line and found that as he paid it out, it was carried first to the south then to the north. He then took water samples and analysed them for salinity; the results enabled him to show that the upper stream was formed of nearly fresh water from the Black Sea and that it retained its identity from, and was compensated by, the deeper, saltier current flowing from the Mediterranean through the Sea of Marmara.

This method of working—questioning local fishermen then amplifying their reports by his own simple experiments—was followed by Marsigli in his three

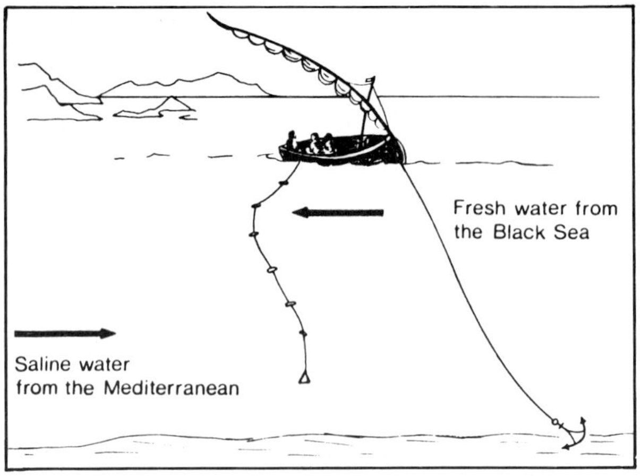

Figure 10. Marsigli's method of measuring sub-surface currents in the Bosphorus.

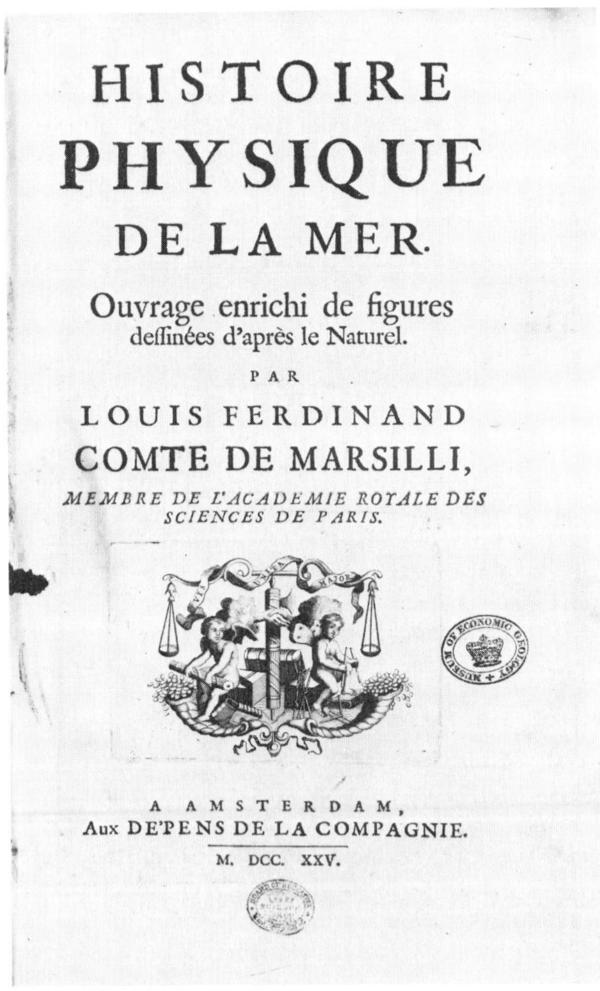
HISTOIRE
PHYSIQUE
DE LA MER.
Ouvrage enrichi de figures
deſſinées d'après le Naturel.
PAR
LOUIS FERDINAND
COMTE DE MARSILLI,
MEMBRE DE L'ACADEMIE ROYALE DES SCIENCES DE PARIS.

A AMSTERDAM,
Aux DE'PENS DE LA COMPAGNIE.
M. DCC. XXV.

Figure 11. Title page of Marsigli's *Histoire Physique de la Mer* (1725).

subsequent hydrographic studies on the Danube, on the Mediterranean coast of Provence, and in Lake Garda.[24]

Marsigli led an active and varied life. After leaving Constantinople he followed a military career fighting against the Turks, during which he rose to the rank of general, only to be demoted in disgrace in 1704 after the surrender of the beseiged town of Brisach, where he was in joint command. During his army service he took every opportunity to study nature, and his mammoth study of the Danube and its region was made at this time. He corresponded with other natural scientists in several countries and in 1691, having been proposed by Isaac Newton, was elected a Fellow of the Royal Society.

Moving to Provence after his demotion, he undertook an intensive study of a small area off the Mediterranean coast in order to expand his theory that mountain ranges did not terminate at the shoreline but dictated the adjacent sea bed topography. His practical work in this area was carried out in 1706–7 and the results were written up by 1708 in the form of a treatise on the natural history of the sea, complete except for a final section which was to deal with the generation of marine animals. The four sections, bound in two folio volumes, were presented to the Royal Academy in Paris and were gratefully received by their Secretary, Fontenel.[25] Meanwhile, Marsigli had been summoned by the Pope to take command of the Papal armed forces and so he returned to Rome taking with him the incomplete manuscript of the fifth and final section of his book. Not until 1725, at the urging of his friend Boerhaave, was the work rewritten and expanded with Boerhaave's assistance and published in Amsterdam as *Histoire Physique de la Mer* (figure 11).[26]

Marsigli's book may be said to mark the beginning of oceanography as we understand the term today, although the name itself did not come into general use until the 1880s. It was based mainly on his work in Provence, with a few additional remarks on his experiences elsewhere. In his preface Marsigli criticised Boyle: although he had published works supposedly about the sea, the contents of Boyle's *Tracts* did not live up to their titles, being based on observations made at second hand. Marsigli accepted that the practical difficulties were indeed great, so that many able men were prevented from completing the work that they had set out to accomplish. He himself, lacking patronage, had perforce been limited to a small area, although this restriction had in fact led to his success. Whereas

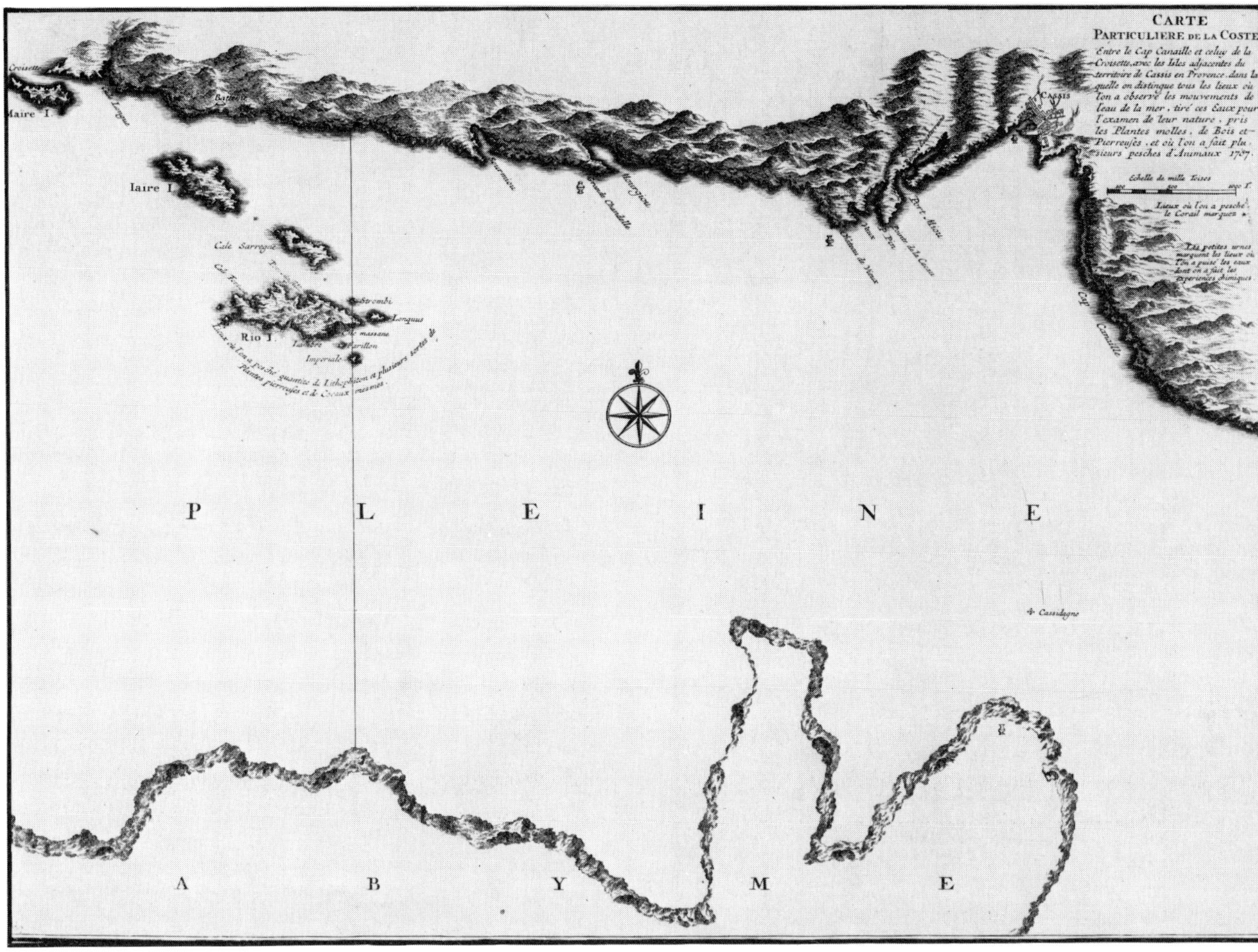

Figure 12. Marsigli's plan of the coast of Provence, showing the location of his section and water samples. From Marsigli 1725 *Histoire Physique de la Mer,* plate 2.

Boyle and Hooke had gathered data from trials in shallow water or undertaken as random stabs into the various oceans, Marsigli worked in the Mediterranean, which is a small though true ocean, its narrow continental margin plunging down to the abyssal basin. He progressed by limited steps, using the same apparatus, and was thus able to determine a pattern of change in the sea bed, its sediments, and the sea water as he moved out from the shore (figure 12).

Marsigli described only his thermometer (figure 13), hydrometer and biological collecting gear. He was able to sound no deeper than 120 fathoms, collecting water samples and temperature measurements to the same depths. His sampler was sufficiently reliable for him to distinguish between the near-surface fresher water diluted by the nearby River Rhône outflow, and the deeper, saltier water.

The thermometer was described thus (in translation):

> I made use of the thermometer whose size and division into degrees is indicated on the drawing placed against the table of these observations. I suspended the thermometer by a line with an appropriate weight, and I lowered it at various places and to various depths during December, January, March and April. I found that the temperature at a depth of 10, 20, 30, 120 *brasses*† stood throughout at 10½ or 10¾ degrees.[27]

† 1 brasse = 63.93 inches or 162 cm.

TABLE

des Experiences faites avec le Thermometre dans la mer à differentes profondeurs et en divers temps et lieux pour examiner la temperature qui s'y trouve.

Lieux où on a fait les experiences.	Années.	Mois.	Jours.	Heures.	État du Thermometre sur l'eau.	Profondeur de l'eau. Brasses	État du Thermometre dans la Profondeur de la mer.
A la Grande chandelle où on a pesché du Corail premiere decouverte des fleurs.	1706	Decembre	7	10 m.	9 ½	10	10 ½
Sur l'abisme à 3. milles au Sud de Cassidagne	1707	Janvier	18	10. m.	8 ½	120	10 ½
A la pointe de l'Est de Rio.	1707	Janvier	26	12 midy	9	20	10 ½
Au petit Rocher nommé strombi à l'Est de l'Isle de Rio	1707	Mars	28	10. m.	12	26	10 ½
A la Calanque massane au Sud-Est de l'Isle de Rio	1707	Mars	29	8	9	30	10 ½
A Lambre au Sud de l'Isle de Rio où on a fait la belle pesche de Corail.	1707	Avril	2	9	11 ½	18	10 ½
A la Coraillade dans l'abyme.	1707	Juin	30	4	15	100	13
Au mesme endroit à la Coraillade	1707	Juin	30	6	15	120	15
Encore au mesme endroit	1707	Juin	30	9	17	1½	17
Le Thermometre s'étant cassé par accident, on n'a pas pû faire d'autres observations qu'on s'étoit proposées.							

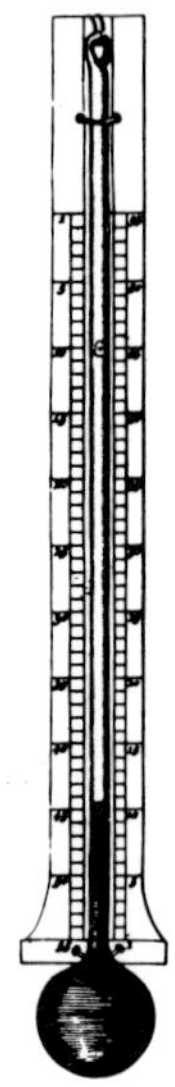

Figure 13. (Left) Marsigli's thermometer, shown beside his tabulated observations. From Marsigli 1725 *Histoire Physique de la Mer,* plate 6.

Marsigli did not indicate whether his thermometer was in any way insulated, perhaps by being secured within his water sampler, but from the data it appears to have been affected by pressure. When he attempted to repeat these measurements the following summer, his boat was attacked by pirates and in the ensuing scuffle the thermometer was broken, so that he could make no further measurements.

The instruments and method of work of Marsigli seem far in advance of those proposed by Hooke only a few years earlier. With careful use of illustrations, maps and sections Marsigli gave a credible picture of the marine environment as he had found it (figure 14). The survey of Lake Garda in northern Italy, which Marsigli undertook in later years, was carried out in the same rigorous fashion.

Figure 14. (Right) Marsigli's profiles of the sea bed off the coast of Provence. From Marsigli 1725 *Histoire Physique de la Mer,* plate 4.

2.4. STEPHEN HALES: NEW IDEAS FROM THE CHEMISTRY LABORATORY

Marsigli's *Histoire Physique de la Mer* was certainly read by the Reverend Stephen Hales (1677–1761), whose pastoral duties evidently allowed him time to pursue his scientific interests. He is now chiefly remembered for his studies of plant and animal physiology for which he too was elected Fellow of the Royal Society in 1717.[28] There is no evidence that Hales ever went to sea, but as a trustee of the colony of Georgia he often met with ship-owners who transported goods and slaves across the Atlantic. Hales tackled the problem of improving the ventilation in ships to prevent the spread of disease, and this brought him to the attention of the Admiralty. Thus encouraged, he made further experiments into the distillation of potable water and the preservation of food at sea. In his account of this work he frequently referred to Marsigli's analyses of sea water. He also attempted over a number of years to construct a sounder and a water sampler.

In 1728 the Fellows of the Royal Society watched a demonstration given by its Curator, J T Desaguliers (1638–1744), of a sounder contrived by Hales and himself (figure 15):

> My contrivance is . . . a Machine which will carry down Mr Hales' Gage to the bottom of the Sea, and immediately bring it up again.[29]

Desaguliers acknowledged the defects of the simple timed sounder and claimed that only a pressure gauge such as that of Hales indicated true depth, irrespective of the time it remained beneath the surface.

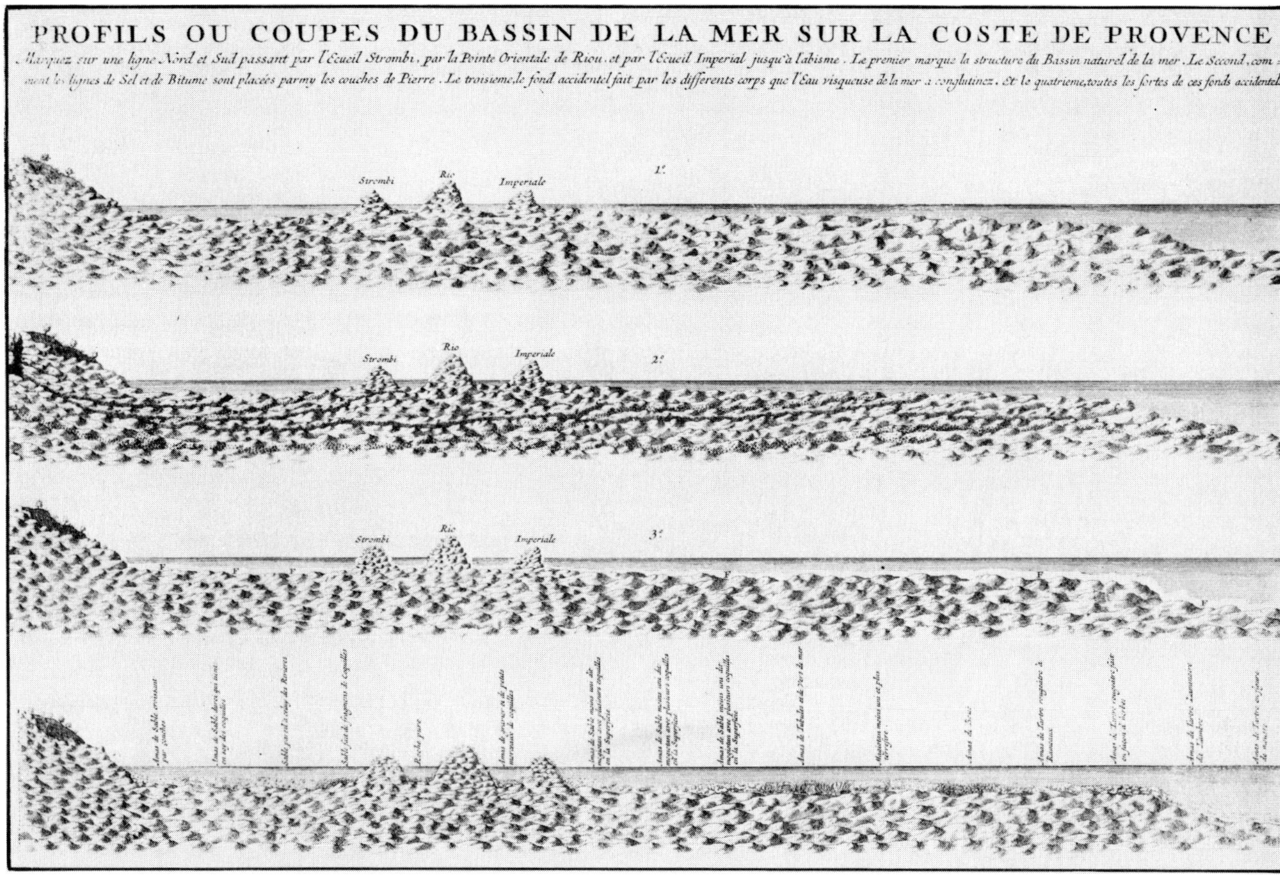

This 'Sea Gage', as Hales later explained,[30] was adapted from one with which he had measured the pressure exerted by a container of peas put to soak (figure 16). It consisted of a vacuum tube inverted over a bottle containing mercury on top of which treacle or honey was floated. The bottle was perforated and increasing water pressure forced the mercury up the tube; when it fell, treacle adhered to the tube marking the highest point reached. Desaguliers had also supplied a detaching sinker mechanism and a bladder float to return the gauge to the surface. With the aid of a pressure tank he showed the effect at a simulated depth of 40 feet.

In a later essay[31] Hales expanded on the practical details of constructing a sea gauge. He suggested replacing the glass tube with one of iron or copper, such as a musket barrel. No mercury was needed, because pressure could operate directly on the contained air. Coloured oil would mark a rod, and this had to be removable since the tube was no longer transparent. The apparatus would be calibrated by lowering it on a line 33 feet into the sea, which would compress the contained air to half its volume. Lowering another 33 feet compressed it to one third, and so on. As the graduations on the marker rod became closer together at each stage, a point would be reached, about half a mile down, when they became indistinguishable. The range could however be extended by adding a metal spheroid whose volume was nine times that of the tube; such a gauge could then be read to depths of 2½ miles.

At this stage Hales had to consider how deep the sea might be, for if the iron tube became porous under extreme pressure then he would have to fit a glass liner in it. There would certainly be difficulties in making pressure-tight joints between the different materials making up the gauge. The weight of the sinker, secured by a spring clip so as to detach at the bottom, and the nature of the float, all depended on the depth at

Figure 15. Pressure gauge devised by Desaguliers and Hales, 1728. From *Phil. Trans.* **35** (405) unnumbered plate.

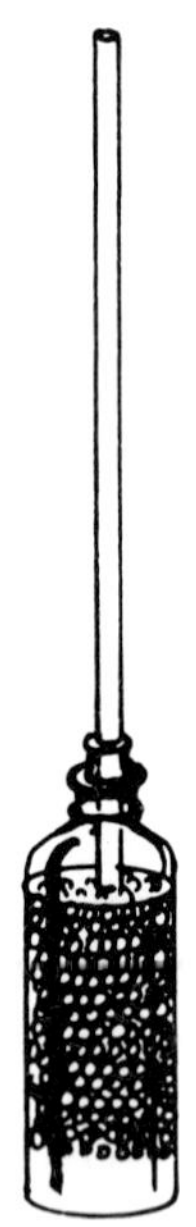

Figure 16. The experiment that gave Hales the idea of a sea gauge. From Hales 1769 *Statical Essays* Fig. 36.

which they were intended to function. Hales speculated that 2½ miles might be the greatest depth of the sea:[32]

> That the sea is not many miles deep is probable from hence, that all the great Oceans are here and there interspersed with Islands; an argument that though as far as the sounding Line has reached, the Sea is found to be deeper and deeper, the further from the Shore . . . which would come to a great depth if indeed it continued on so . . . but the interspersed Islands prove that it is not so.

He acknowledged, on the other hand:

> If we suppose the Sea to deepen from Shores in nearly the same proportion that Land rises from the Shores, then from the following Estimate the greatest Depth of the Sea will not exceed five or six Miles.

Carrying this reasoning further, if the volume of sea water equalled the volume of land above sea level, since oceans occupied so great a proportion of the globe, they must be considerably less deep than the average height of the land. If his gauge accelerated during the descent, Hales thought that it might strike the sea bed with great force and suffer damage. He recommended that trial runs with a metal dummy and float should be made or, alternatively, that the sinker should be carried some distance below the gauge to lessen the impact.

Like so many contemporary devices, Hales' gauge started out as a basically simple concept but was modified to such an extent that it became unworkable. The essay was reprinted in *Gentleman's Magazine* in 1754[32] with an illustration (figure 17) and an appendix stating that a gauge had been made by Francis Hauksbee, well known instrument maker, and had answered well on trials on the River Thames. A similar instrument used by Erasmus King in the Gulf of Finland had registered a depth of 288 feet which was confirmed by use of lead and line. However, Hauksbee's gauge was tested by Colin Campbell near the Bermudas:

> . . . and on a clear day was let down into the sea . . . several ships being in company, and a good lookout ordered from them all; yet it was not seen to return, though they waited for it between three and four hours.[32]

In 1751 a letter from Captain Henry Ellis addressed to Hales was read to the Royal Society,[33] describing trials made on the *Earl of Hallifax* with a Hales ventilator and a 'bucket sea-gage'. This latter apparatus, which Hales described in a letter to the Society's President in reply to that of Ellis, seems never to have

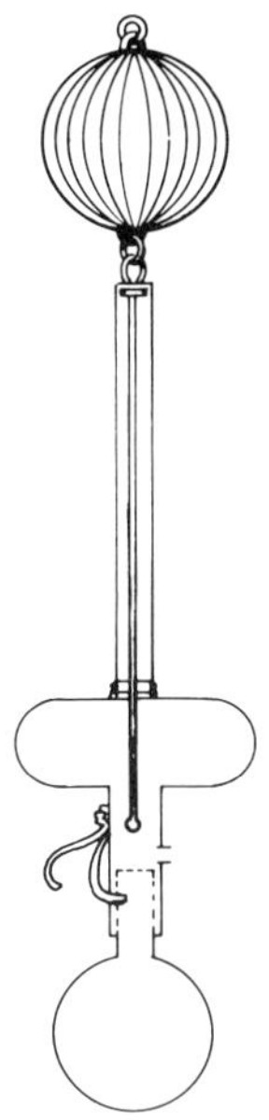

Figure 17. Hales' 'improved sea-gage', 1754. From *Gentleman's Magazine* **24**, Fig. 11.

been illustrated, perhaps because of its very simplicity. Hales wrote:

> The bucket sea-gage, which he mentions, and which I provided for him, to find the different degrees of coolness and saltness of the sea, at different depths, was a common household pail or bucket, with two heads in it; which heads each had a round hole in the middle, near four inches diameter, which were covered with valves which opened upwards; and that they might both open and shut together, there was a small iron rod fixed in the upper part of the lower valve, and at the other end to the under part of the upper valve; so that both the valves opened by the force of the water, which had by that means a free passage through the bucket. But when the bucket was drawn up, then both the valves were shut by the force of the water at the upper part of the bucket: by which means the bucket was brought up full of the lowest sea-water, to which it had descended.

A lead weight was slung below the bucket to carry it down vertically. The bucket also carried a thermometer to be read when it was brought back on deck:

> When the bucket was drawn up, the hole at the bottom was stopped with a cork, to keep the water in, when the valves were opened, to come at the mercurial thermometer, which being tied to an upright stick, could readily be unfastened, by pulling out a loose nail, which went into the upper end of a stick, which was fastened at its lower end in the same manner.

This thermometer was not self-registering and had therefore to be read before its bulb was lifted out of the water.

Ellis made several trials, letting the bucket down to between 360 and 5346 feet, using a small Fahrenheit thermometer made by Bird. He found that the coldness increased regularly down to 3900 feet and remained constant at greater depths. He thought that the water had warmed slightly on its way up as the surface temperature at the time was 84 °F. He took a great quantity of line for deep soundings, but did not have with him a Hales depth gauge.

2.5. COCK AND DOUGLAS: APPARATUS MADE ON BOARD SHIP

Two other descriptions of apparatus appeared in the *Philosophical Transactions* during this period. A 'Machine for sounding the Sea at any Depth', which was merely a sophisticated version of that proposed by Hooke, was invented by Major William Cock during a voyage to Georgia in 1738 (figure 18).[34] It differed from earlier versions in that its detachable weight was carried on the main shaft of the sounder, rather than at its base. This sounder would suffer from the same disability as all other wooden devices: the tendency to become waterlogged and the problem of spotting the float when it returned to the surface.

The second article[35] described a carrying-case for a thermometer, which Captain Charles Douglas had lowered to 260 fathoms from his ship HMS *Emerald* off the coast of Norway in 1769. Douglas used a tin cylinder of a quart capacity, filled with water and sealed with the thermometer fixed within it. Since the vessel insulated the thermometer, the tin had to be left hanging for a considerable time, and so to avoid this Douglas made two small holes in the ends and sent it down empty, so that it would fill below the surface. He then hauled it up as rapidly as possible, finding that it took 13½ minutes to raise from 260 fathoms.

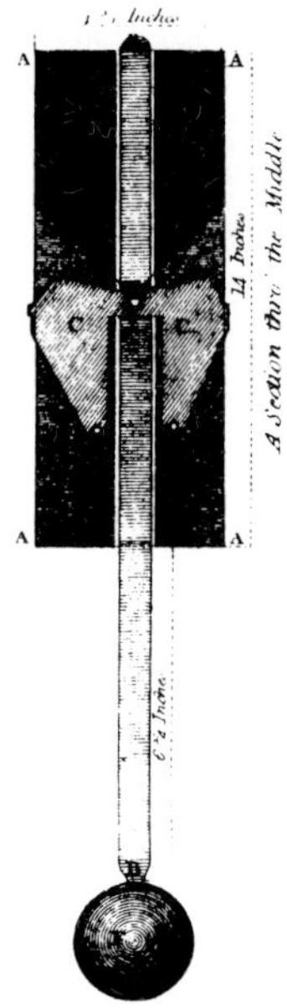

Figure 18. William Cock's sounder. AAAA is a square timber; BB is a transverse timber grooved to carry CC, two pieces of iron whose feet rest on DD, two pins passing through the trunk. E is an iron pin carrying the upper ends of CC. F is a weight. From *Phil. Trans.* (Abridgements) 1744–49, plate 5, Fig. 1.

Douglas' experiments showed correctly that during May the surface water was several degrees colder than that at depth, whereas by July the difference was less pronounced. With holes pierced at each end, the contents of the container would have been steadily diluted on the way up, as well as being modified by the ambient water temperature during the relatively slow ascent. For his deepest sounding Douglas gave these figures: air 46°F; surface water 47°F; 100 fathoms 46°F; 260 fathoms (not at the bottom) 52°F.

REFERENCES AND NOTES

1. Boyle R 1671 *Tracts on the Cosmicall Quality of Things* (Oxford)
2. *Op. cit.* pp4–5
3. *Op. cit.* p7
4. *Op. cit.* p7
5. *Op. cit.* p8
6. *Op. cit.* p13
7. *Op. cit.* pp13–14
8. *Op. cit.* p5
9. Royal Society, *Hooke Mss XX No.* 23
10. Anon 1666 'An Appendix to the Directions for Seamen Bound for Far Voyages' *Phil. Trans.* **1**(9)
11. *Op. cit.* pp147–8
12. Multhauf R 1960 'The lineless sounder: an episode in the history of scientific instruments' *J. Hist. Med.* **15** pp390–8. For an accessible translation of one such sounder, see Dolan J 1962 *Unity and Reform: Selected Writings of Nicholas de Cusa* (Indiana: Notre Dame) p253
13. This brief note was printed in *Phil. Trans.* **1** (1666) p178
14. *J. Sçavans* 1666 **1**(18) pp419–23
15. *J. Sçavans* 1666 **1**(21) pp457–63
16. Royal Society, *Hooke Mss XX No.* 35
17. 'Some considerations touching a letter in the Journal des Sçavans of May 24, 1666' *Phil. Trans.* **1** pp228–30
18. If made of brass this sampler may have been cylindrical, in which case, as experiment will show, the lower lid cannot open far into the base.
19. Hooke R 1667 'Directions for observations and experiments . . .' *Phil. Trans.* **1** pp433–48
20. These 'Cutlerian Lectures' were later published with illustrations as part of *The Curious Philosophical Observations and Experiments of the Late Dr Robert Hooke* 1726 ed. W Dereham (London) pp225–48
21. Nicholson W 1772 *An Introduction to Natural Philosophy* 2 vols (Edinburgh) vol 2 pp106–8
22. His name sometimes appears in the French spelling as Louis-Ferdinand de Marsilli
23. Marsigli L-F 1681 'Osservazioni intorno al Bosforo Tracio o vero Canale di Costantinopoli rappresentate in lettera alla Sacra Real Maesta di Cristina Regina di Svesia da Luigi Ferdinando Marsili' (Rome); Reprinted in *Boll. Pesca, Piscicoltura e Idrobiologia* 1935, Ao XI(5) pp734–58
24. Marsigli L-F 1726 *Danubius Pannonico-mysicus* . . . 6 vols (Amsterdam), ref (26), and 'Osservazioni fisiche intorno al lago di Garda', in Sorbelli A (ed) 1930 *Scritti inediti . . . raccolti e publicati nel 2° centenario dalla morte a cura del Comitato Marsiliano* (Bologna)
25. Marsigli L-F 1711 *Osservazioni Naturali Intorno al Mare* (Venice)
26. Marsigli L-F 1725 *Histoire Physique de la Mer* (Amsterdam)
27. Marsigli's thermometer predated the invention of the Réamur scale and it was probably divided arbitrarily, which accounts for his having abandoned observations once the instrument was lost, since he could not have obtained a replacement thermometer with the same scale markings.
28. Clark-Kennedy A E 1929 *Stephen Hales, DD, FRS, an Eighteenth-century Biography* (Cambridge)
29. Desaguliers J T 1728 'An account of a machine for measuring any depth in the sea with great expedition and certainty' *Phil. Trans.* **35** pp559-62
30. Hales S 1769 *Statical Essays* vol. 1 (London) p209. This reference is to the fourth edition of this work, the only one available to the author.

31. Hales S 1733 *Statical Essays* vol.2 (London) pp332–47
32. Hales S 1754 'A description of a sea gage to measure unfathomable depths' *Gentleman's Mag.* **24** pp215–19
33. Ellis W 1751 'A letter to the Rev. Dr Hales FRS from Captain Ellis FRS dated 7 Jan 1750–51 at Cape Monte Africa, ship "Earl of Hallifax" ' *Phil. Trans.* 47 pp211–14, followed by Hales' letter, pp214–16.
34. Cock W 1746 'A machine for sounding the sea at any depth or in any part invented by Major Wm Cock in the year 1738, in a voyage to Georgia' *Phil. Trans.* **44** pp146–7
35. Douglas C 1769 'On the result of some attempts to ascertain the temperature of the sea in great depths, near to the coasts of Lapland and Norway by Charles Douglas FRS, then captain of HMS "The Emerald" Anno 1769' *Phil. Trans.* **60** pp39–43

Chapter 3

From Amateur to Professional Instrument-makers

3.1. DEEP-SEA THERMOMETERS OF CAVENDISH AND DE SAUSSURE

Any marine thermometer must be capable of preserving the temperature registered below the surface while being hauled up. The thermometer may register extremes of temperature, or that of a pre-selected level.

Prior to the nineteenth century these alternatives were seldom appreciated and it was commonly supposed that temperature simply decreased with depth. This holds broadly true, but there are exceptions, notably in polar seas where temperature profiles may be distorted when cold, fresh meltwater overlies warmer but more dense saline water.

In 1757 Lord Charles Cavendish (1703–83), Vice-President of the Royal Society, described a set of registering thermometers,[1] all of the overflow type (figure 19), in which a large volume of thermometric fluid, usually alcohol, drove a smaller volume, usually mercury, into an overflow chamber. When the temperature returned to its former level the mercury trapped in the chamber indicated the extreme temperature that had occurred. Cavendish suggested that such thermometers might serve for marine uses:

> If instruments . . . were to be used for finding the temper of the sea at great depths, some alteration would be necessary in the construction of them, principally on account of the great pressure of the water; the ill effect of which can, I believe, be prevented in no other way, than by leaving the tube open. For if the thermometer was made strong enough to resist the pressure without breaking, yet it would be impossible to be sure, that the figure of the glass would thereby be altered, which should make the experiment uncertain.
>
> The instrument for finding the greatest heat might be made just like that of figure 1 only leaving the top open. It is to be filled with mercury only, as is also the lower part of the ball at top, but not near so high as the end of the capillary tube. The upper end of that ball . . . will in a great measure be filled with the sea-water, which will be forced into it by the pressure.
>
> If this instrument (the tube being quite full of mercury) is plunged into any part of the sea, where the heat is greater than that of the air above, part of the mercury will be driven out of the tube; and upon bringing it into a colder place, the sea-water or air in the ball will enter into the tube, and will fill the space left by the mercury.
>
> As this thermometer does not show the common degrees of heat, it must be placed in a vessel of water with another thermometer, and the scale of degrees at top will show how much the heat it has been exposed to is greater than that of the water in the vessel.
>
> The sea-water getting into the glass will corrode the mercury, and thereby foul the glass, which will make the experiment less exact.

This could be avoided, noted Cavendish, by tying a bladder full of air over the tube end and in this case both mercury and spirit could be used, but the advantage with mercury alone was that the instrument could be made smaller overall.

The minimum thermometer shown in Cavendish's Fig. 3 would also have to be opened, and so he proposed reshaping the tube to that of his Fig. 4, which was better suited to withstand a rough passage and more easily reset. To find the absolute temperature:

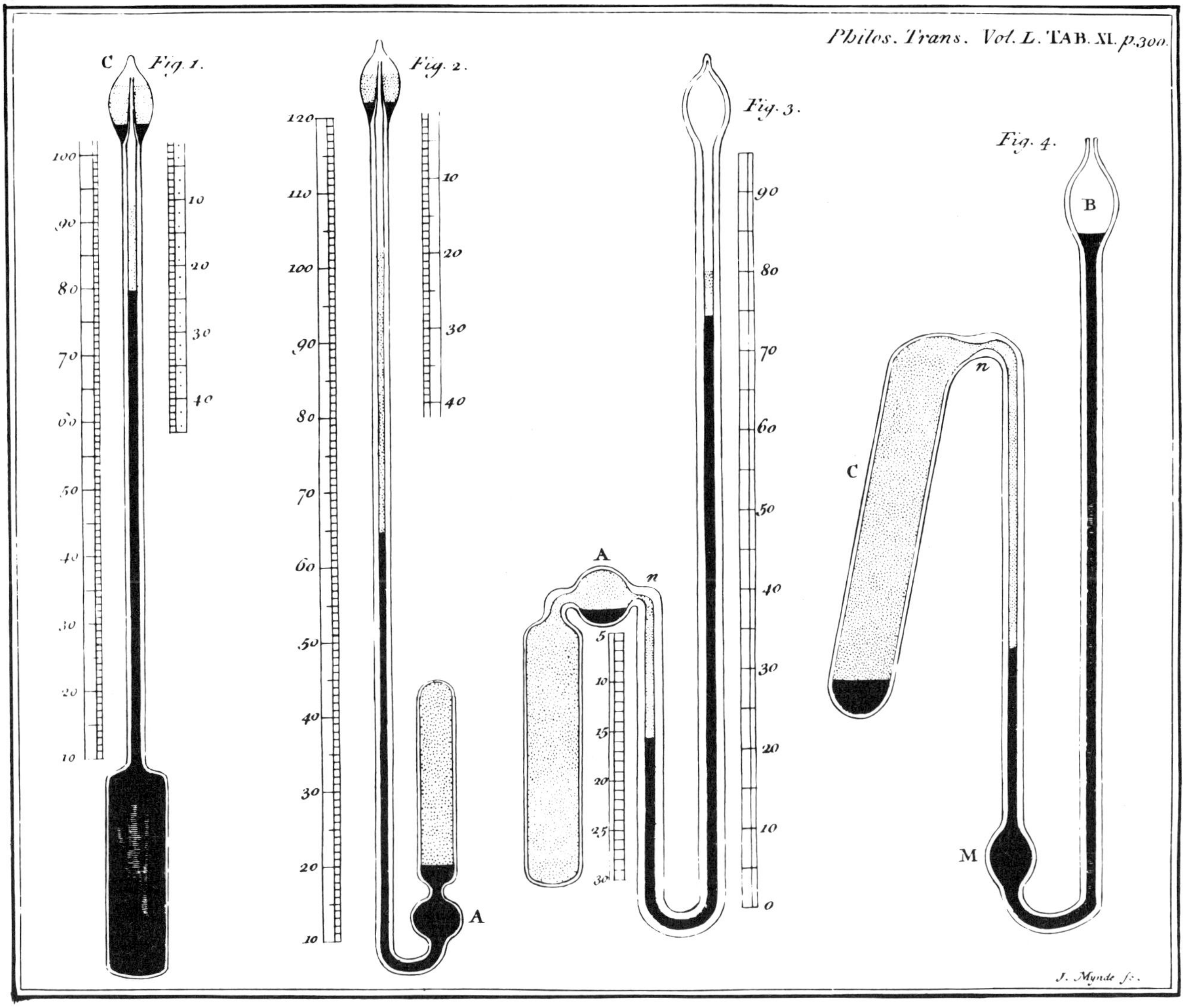

Figure 19. Cavendish's overflow thermometers, 1757. *Right:* deep-sea form. From *Phil. Trans.* **50**, plate 40.

> If, by means of a bladder, the sea-water can be kept out of the glass, this instrument may be made to show the common degrees of heat; but even then, in order to render it less bulky, it may be better to supply the want of them as in the last instrument. The longer leg of the syphon may in that case be made as short as you please; only making the ball B big enough to receive all the mercury, which may be driven into it by heat.

The constriction *n* reduced the size of mercury droplets passing through, thus enhancing the sensitivity of registration. The Cavendish minimum thermometer was taken on Phipps' Arctic expedition of 1773 and the precautions taken to guard against the distorting effects of pressure are recounted in §4.1.

Overflow thermometers were crude instruments, inevitably registering in steps as each mercury droplet passed into the overflow chamber. For greater precision, Horace-Benedict de Saussure (1740–99), professor of philosophy at Geneva, chose instead to insulate a thermometer so that it could be lowered into the water and left to adjust to its surroundings, holding its registration unchanged whilst it was quickly raised to the surface. Whereas Cavendish's thermometers registered extremes of temperature, de Saussure's would register at a known level. In his book *Voyages dans les Alpes*[2] de Saussure described his researches on temperature within the sea, in lakes and in the Earth.

In contrast with the thermometers used in meteorology, where maximum sensitivity was required, de Saussure's model was made of thick glass with a one-inch diameter bulb filled with alcohol, which is a slower conductor than mercury. He made up some soft wax mixed with oil and resin, and moulded a three-inch casing around the thermometer base, putting this into a wooden box and securing the whole assembly with a metal hoop so that the wax penetrated every space and joint. Only the graduated stem of the thermometer projected, shielded by a grille of metal rods spreading from the suspension ring (figure 20). The whole apparatus was weighted by a lead sinker.

After successful shallow-water trials, de Saussure moved to a location off the coast of Provence where he sounded in 1918 feet of water. He left the thermometers down all night, attached to a buoy. When hauled up, the insulated thermometer stood at 10.6° Réamur; its companion, by the same instrument-maker but non-insulated, had been crushed by the pressure.

From his previous experiments de Saussure knew of the length of time required for his thermometer to take up the temperature of its surroundings, and by how much its reading would alter during its 24-minute ascent. Although the result was satisfactory, the period required for it to adjust to its surrounding temperature prevented its regular use at sea. Near the coast it could be hung from a buoy, but in the open ocean no ship could expect to remain hove-to for many hours except in a dead calm and in the absence of currents.

Fig. 3.

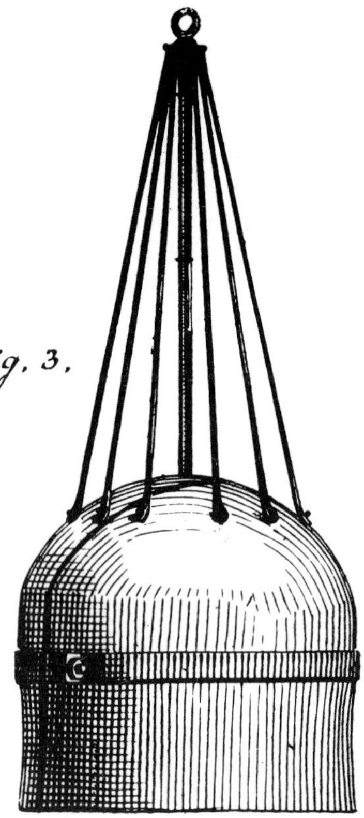

Figure 20. De Saussure's insulated thermometer. From de Saussure 1797 *Voyages dans les Alpes* vol. 3 plate 2, Fig. 3.

3.2. THE SIX SELF-REGISTERING THERMOMETER (1780–93)

James Six (1731–93) was descended from one of the Hugenot families who arrived in Canterbury during the reign of Elizabeth I. He had been occupied in the family silk-weaving business but retired early to devote himself to the natural sciences. He had the catholic range of interests typical of an eighteenth-century intellectual, corresponding and exchanging observations with astronomers, and he also built a number of electrical machines for the medical profession. He is best known for the invention of the self-registering thermometer, still popular today, which bears his name. He wrote several papers on the instrument[4] and was elected Fellow of the Royal Society in 1792. Later he wrote a book on the thermometer[5] which was published in 1794, a year after his death.[6]

The self-registering thermometer was not a new concept—Hooke had proposed a marine minimum thermometer (§2.2), and in 1698 Jean Bernoulli had described in a letter to Leibnitz one which might register both maximum and minimum temperatures.[7] Bernoulli's thermometer used air as a thermometric fluid (but this is unsatisfactory since it responds to changes in barometric pressure as much as to temperature), while its tube, designed to trap and retain some of the registering fluid, was corrugated in a way that was beyond the glass-blowers' competence to manufacture at that time. The Cavendish thermometer, though more efficient and of simpler construction, was fragile, difficult to use and set, and somewhat imprecise.

Six's design was fundamentally different, and it worked in the following manner: in a tube folded into three limbs, alcohol filled the central expanded reservoir (figure 21, *left*) and the upper portion of the left-hand limb, separated by a mercury sector occupying the U-bend. The liquids could expand into the upper corked chamber and were kept in contact with each other by atmospheric pressure. In each outer limb light steel-in-glass indexes floated in the alcohol; these could be pushed up by the mercury meniscus but were prevented from falling down by slender glass 'tails' acting as springs.

As the temperature rose, the alcohol in the central reservoir expanded, pushing the mercury and index up the left-hand limb. Any subsequent drop in temperature caused the alcohol to contract; the mercury receded, the left-hand index remained to record the maximum, while the right-hand index was lifted to record the lowest temperature. After use the indexes were repositioned at the meniscus with the aid of a magnet.

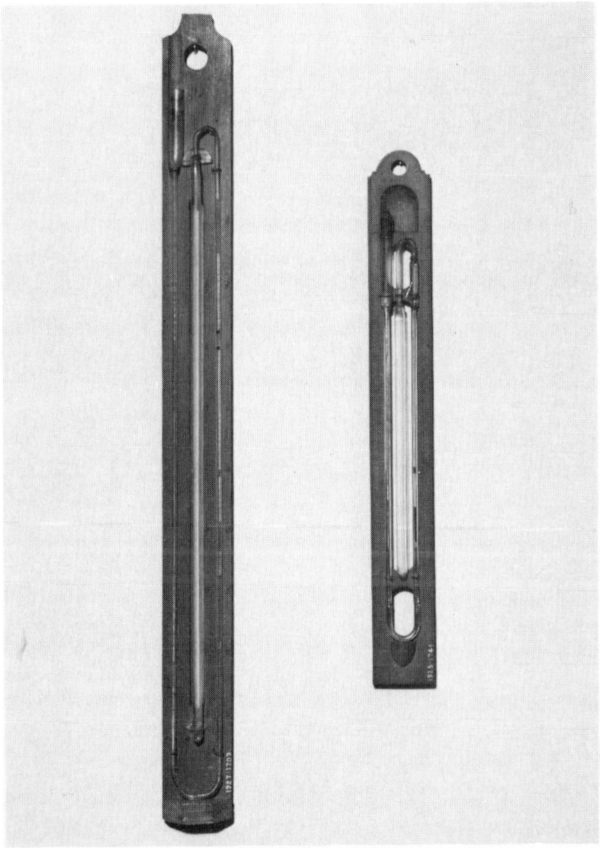

Figure 21. Original maximum and minimum thermometers by Six, *c*1780. *Left*, meteorological thermometer; *right*, deep-sea thermometer.

Six devoted a chapter of his book to a description of deep-sea thermometers.[8] He considered that none of those available had been satisfactory, recounting the measurements of Douglas (see §2.5), of Forster during Cook's voyage of 1772–73 (probably made with a Hales bucket), by Phipps in 1773 (see §3.1), and by de Saussure (see §3.1). The last of these was probably the most accurate but it was impractical for use at sea. The instrument proposed by Six was shorter and of thicker glass than his meteorological thermometers. The indexes were steel needles sealed into glass tubes, with a button or 'bugle' at each end replacing the spring tail. When not in use, these indexes were to be stored in wide-bore upper sections of the tube (figure 21, *right*).

At sea, the thermometer was first brought to the temperature of the surface water and the indexes moved down, to the surface of the mercury with a magnet. As the thermometer sank into the depths, the mercury in one limb would descend, drawing an index down with it. On hauling in, the rising mercury would pass beside the index, forcing it against the wall of the tube, where it remained to mark the extreme temperature reached. 'Maximum' and 'minimum' indexes would therefore be found on opposite sides to those of the meterological thermometer. Six pointed out that indexes of this type, although a liability in a meterological thermometer, had an advantage in deep-sea use for, once immersed in the mercury, no amount of shaking could dislodge them. Indeed, they could not even be shifted with a magnet, and to reset the instrument it often had to be cooled (by the application of ether to the bulb) below the lowest temperature registered, in order to free the index.

However, despite Six's application to the problem, and his published design, it was the meteorological instrument that proved to be best suited to marine use; there is no evidence that the deep-sea thermometer was ever actually made, other than by Six himself.[9] General Aboville suggested in 1798[10] that an additional sector of mercury might be needed to replace atmospheric pressure if the tube was sealed, but this turned out to be unnecessary. The elasticity of a little air under a sealed cap was adequate to keep the fluids in contact when they contracted. However, the tube was reduced in overall size, and the indexes made shorter. Bristle replaced the glass spring tails and some creosote was mixed with the alcohol to preserve their resilience. In this form it served for the next half century.

The first known use of the Six thermometer at sea was in 1803, on the Imperial Russian Navy ships *Nedezhda* and *Neva*. Captain A J von Krusenstern, who had previously served with the British Navy, was on this occasion exploring a route from Russia's European ports round the world to her eastern shores, and her North American territories. During the passage through the southern hemisphere it was considered worthwhile to undertake geomagnetic and astronomical observations, for the improvement of navigational science. Von Krusenstern's *Introduction* to his narrative listed

> The instruments which Mr Horner [the scientist on board] brought with him from Hamburg, or bought afterwards in England . . . A thermometer, of the kind invented by Six, which shews the greatest degree of heat or cold, that has taken place during the absence of the observer, as a supplement to an instrument for ascertaining the temperature of the sea under water, and which was made by Scheschurin, a Russian artist.[11]

This grammatically confused sentence implies that the thermometer was of English origin and that von Krusenstern also had a 'Hales machine' (as he later termed it) to collect water for subsequent measurement. West of Hawaii, in June 1804, von Krusenstern reported:

> Dr Horner availed himself of the calm to make some observations on the temperature of the water. He found that in Six's thermometer the quicksilver fell 13°3′ (Réamur) at a depth of 125 fathoms; on the surface of the water it stood at 20°5′, so that a difference took place of 7°2′; Hales machine only shewed a difference of 2°5′. At 50 fathoms Six's thermometer was at 17°3′ and at 25 fathoms 19°7′; thus in these latitudes the difference in the temperature of the water at a depth of 25 fathoms is one degree; at 50 fathoms 3°3′; and at 125 fathoms 7°2′ Réamur.[12]

In view of his position at the time, von Krusenstern was probably justified in his assumption that the deepest water was the coldest and therefore giving credit to the greater accuracy of the Six thermometer. From then on the Six thermometer was employed on the majority of scientific voyages together with water-bottles into which registering or non-registering thermometers were fitted. French scientists were less satisfied with the performance of the 'thermographe' as they termed the Six instrument, and often preferred the overflow thermometer, as improved and popularised by Walferdin.

In 1875 the geologist (later Sir) Joseph Prestwich published a tabulation, analysis and discussion of all known temperature measurements taken between 1749 and 1868.[13] Having begun this painstaking survey many years earlier, when some of the persons responsible for the original measurements were still alive, Prestwich had questioned them on the methods used, where this was in doubt, and he also had access to some unpublished results. One of the problems he sought to resolve was that of the protection from pressure given to the Six thermometer during the early decades of its use.

Lowering or raising a fragile glass instrument over the side of a rolling vessel was a hazardous operation and it seems always to have been the practice to put the thermometer into a metal box or tube. If the tube was perforated the thermometer would quickly take up the ambient temperature, but if an attempt was being made to protect it from compression, it would be logical to seal the tube and to leave it down long enough to transmit the surrounding temperature to the thermometer within.

It should have been apparent to its users that the Six thermometer was susceptible to pressure, even if thick glass was used in its construction. The mercury could be moved by the pressure of finger and thumb upon the bulb and at depth would give a falsely high reading. This, together with some ignorance of the precise temperature at which sea water attains maximum density, led to the erroneous conclusion that there was a uniform deep-water mass with a temperature around 40°F. The arguments which were presented for and against this concept will not be discussed here, but Prestwich was able to deduce that some form of protection had been adopted by the French Captain Dumont D'Urville in 1826 and by his compatriot Bérard in 1831.

The first to describe such precautions was Captain Abel Du Petit-Thouars, who sailed in *La Vénus* on a circumnavigation lasting from 1836 to 1839. During the voyage he made 59 attempts to measure sub-surface temperatures, not all of which were successful. His thermometers, by Bunten of Paris, were sealed in strong brass cylinders 15.6 mm thick, and were lowered and left to hang for one hour. When the extreme pressure forced water into the cylinder, a correction was applied according to the depth, which was itself estimated by compensating for the angle of descent of the sounding rope.

Before about 1830, therefore, allowance was not always made for inaccuracy due to pressure and, indeed, any arithmetical compensation depended on the accurate measurement of the depth to which the thermometer had been lowered. Protection did not become routine even after Du Petit-Thouars' results were discussed by François Arago in 1840.[14]

3.3. THE ORIGINS OF THE MASSEY WAYWISER SOUNDER

About 1785 a lineless sounder was invented by an employee of the Board of Ordnance named Greenstreet. He submitted it to the Royal Society for the Encouragement of Arts but they refused it an award until it had been tried out at sea. Unable to comply, Greenstreet later passed the sounder to a shipbuilder, John Charnock, who described its performance in shallow-water trials made in 1795 (figure 22).[15] Reporting that it performed accurately even within strong currents, he claimed that the only disadvantage in its use was the need to lower a boat to recover the sounder when it resurfaced. The dial mechanism was capable of registering 10 miles 56 fathoms, but this was a hopelessly optimistic figure since the greater part of the sounder was made of wood.

This single published example of a waywiser sounder is less isolated than it might appear. The principle of a vane or wheel geared to a pointer moving round a dial is common to clocks, ships' logs and river current meters, and in fact such instruments were all being made at this time. The waywiser sounders made by the first Edward Massey to enter this field drew on such traditions, Massey himself having been trained as a clock-maker.

The clock mechanism consists of all parts positively driven, that is, gears and hands are in physical contact. In similar log mechanisms the log vane and the counter are both in the water and after a timed run are hauled aboard to be read. In others, a rope-drive linked the vane to its ship-board counter. The de Saumarez log of 1725 (figure 23) exemplified this type.[16] For accuracy of measurement each metal part has to be directly coupled to the next, as rope linkage tends to unlay or twist more tightly as the rope turns.

In William Foxon's log of 1772 a towed helical wooden screw turned a line connected to a gear train on board. This log was certainly constructed for examples survive in various collections, but there is no suggestion that it was ever converted for sounding.[17]

The 'Perpetual Log' of Richard Hall Gower's 1792 patent differed in that the helical screw and register were held at the surface of the water within a float.[18]

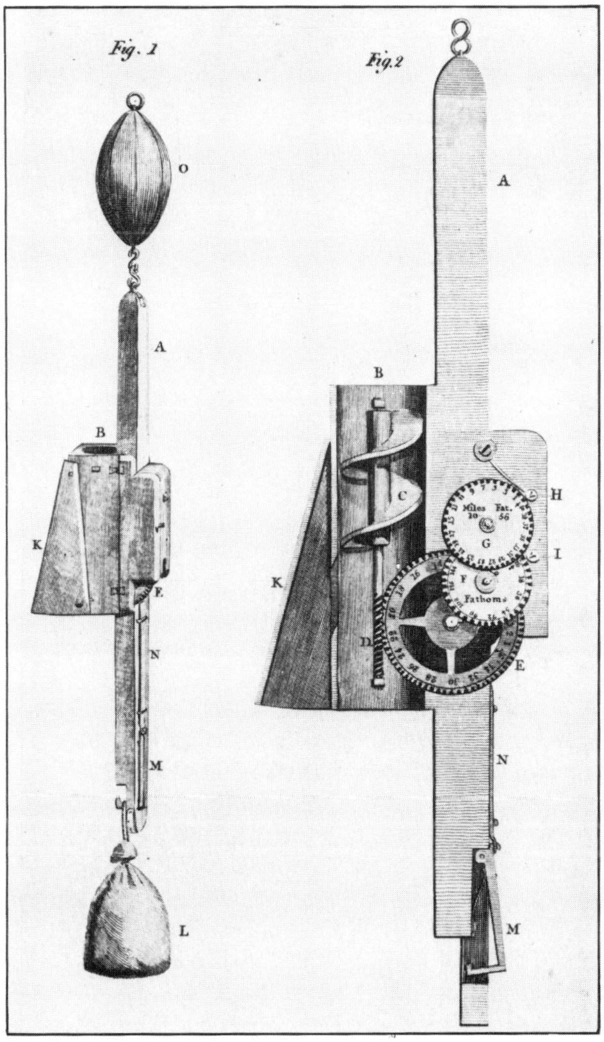

Figure 22. Charnock's illustration of Greenstreet's sounder. A, wooden shaft; C, wooden helix driving gears G, F, E; K, deflector feeding water to the helix; M, detaching hook releasing weight L. Float O carries the sounder back to the surface. From Charnock 1795 *Rep. Arts and Industry* **2**, plate 12.

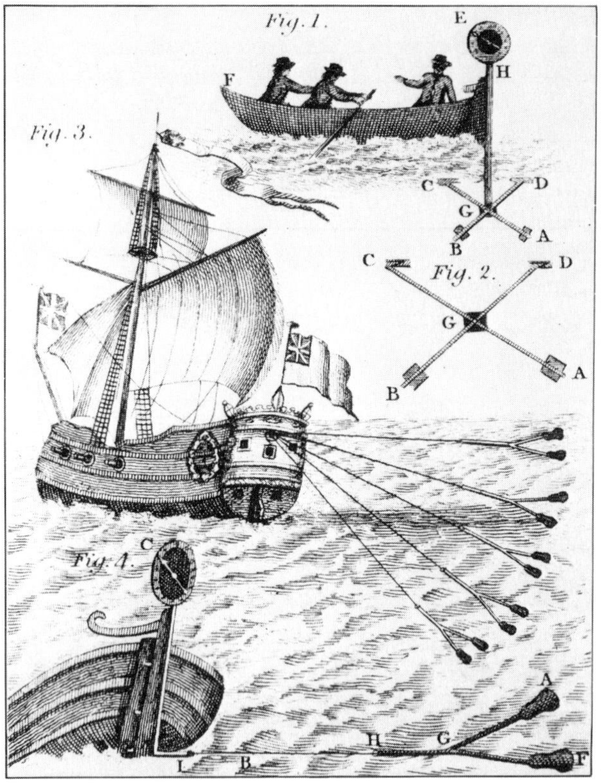

Figure 23. De Saumarez' log, 1725. From *Phil. Trans.* (Abr.) **7**, plate 8.

Chester Gould's log (figure 24), which was first patented in America in 1799, had a four-finned cone or 'fly', and the recording mechanism was kept on board ship. Gould took out two British patents when he moved to London in 1800.[19]

Such was the technical situation when the Massey family of instrument-makers added ship's logs and sounders to the range of clocks that they were already making at the end of the eighteenth century. Several members of the family were named Edward and they also moved from town to town. Figure 25 shows the working lives of those to be mentioned hereafter.[20]

As a craftsman skilled in brasswork and gear-cutting Edward (2) was well qualified to embark on the design of navigational instruments based on the waywiser principle, although there seem to have been no previous seafaring connections in the family. He worked in the Newcastle-under-Lyme area until 1813 when he moved to Coventry and then to Prescot, arriving in London around 1833 where he established himself in Clerkenwell until his death in 1852. He

Figure 24. (Right) Gould's log, 1799. From Gould 1801 *Rep. Arts Manuf.* **15** plate 13.

manufactured large numbers of instruments for the Admiralty and this business was continued by his son Edward (3). Massey's designs were followed by several instrument-makers in the later nineteenth and early twentieth centuries, the firm finally being taken over by Thomas Walker and Sons of Birmingham, who were related to the Massey family. This company continues to the present day.

In 1802 Massey took out a patent for design features to improve the construction of logs, current meters and sounders (figure 26).[21] It was Massey's premise that, when used from a ship under way or carried along by a current, his waywiser mechanism would register only the *vertical* distance travelled and he therefore attached his sounder to a conventional lead and line. The main feature of this patent was an 'air-tube' or hollow shaft which so lightened the vane that it became almost weightless in water. Even when barely moving it was very sensitive and rotated to give an accurate register. (The problem is analogous to constructing an anemometer to register light winds.)

In the patent drawing shown in figure 26, Massey's Fig. 7 showed a device to keep the line perpendicular when sounding from a ship under way. A self-contained reel of cable was carried on a buoyed frame-work. The reel was cast overboard, together with the sounder attached to its rope. The ship then moved

Edward (1)	*b*1740	fl 1767–?1813	*d*1813
Edward (2) Son of Edward (1)	*b*1768	fl 1790–?1830	*d*1852
Edward (3) Son of Edward (2)	*b*1794	fl 1825–1843	*d*1867
Francis Joseph Brother of Edward (3)	*b*1801	fl *c*1824–1841	*d*1845
Thomas Brother of Edward (3)	*b*1803	fl 1828–1865	*d*1868
Edward John Grandson of Edward (1)	*b*1810	fl *c*1834–*c*1860	*d*1872
John Edward Nephew of Edward (3)	*b*1831	fl *c*1865	*d*1891

Figure 25. Some members of the Massey family known to have patented improvements to logs and sounders. From Treherne 1977 *The Massey Family* pp 3–4.

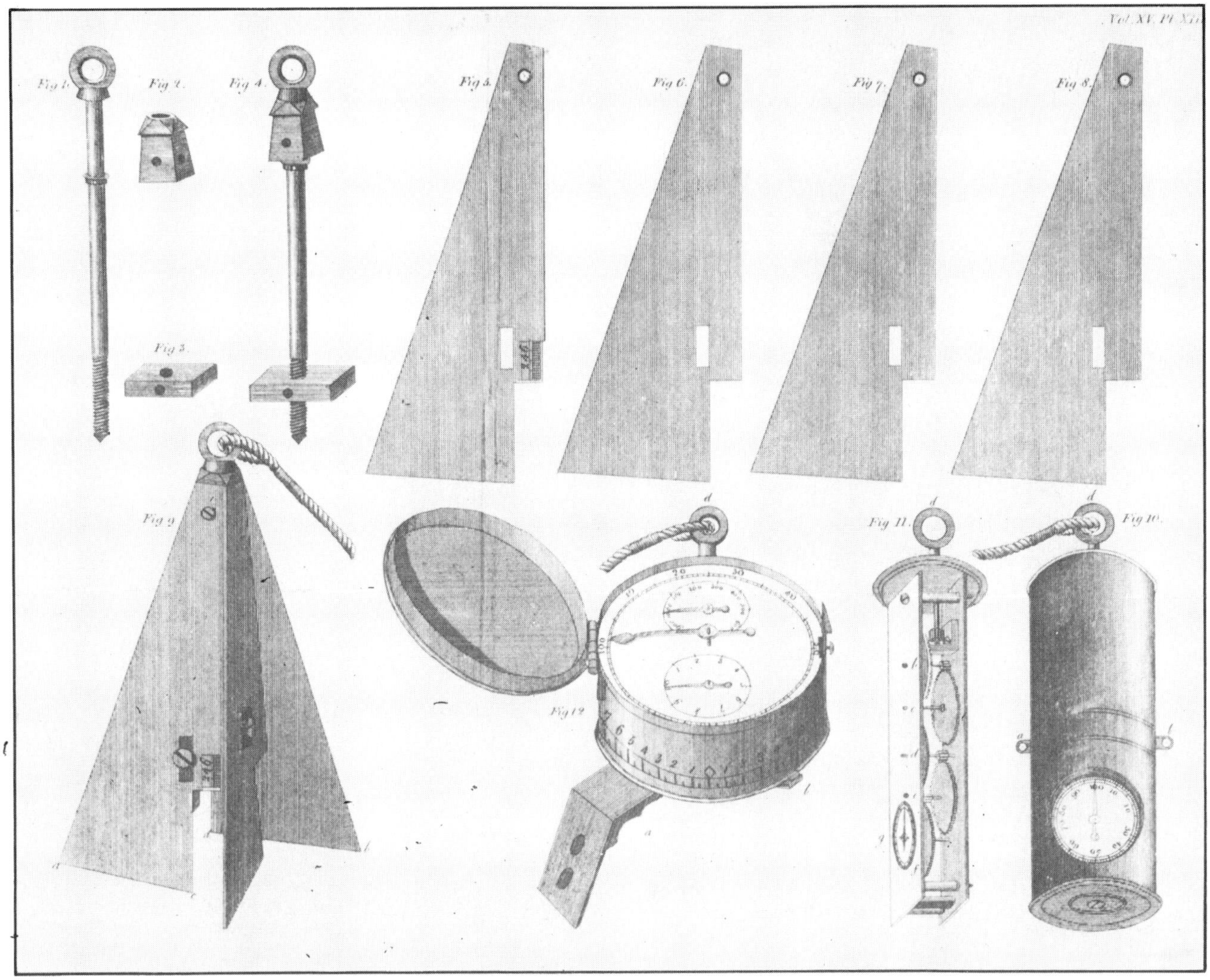

away, paying out sufficient rope to leave the reel over the point at which the sounder had entered the water. When the sounder was deemed to have reached the bottom, a matter which the leadsman was expected to recognise by watching for a decrease in the reel's turning, a tug on the veered rope locked the reel and the entire apparatus was hauled in.

Massey specified that the reel was not intended for soundings which required more than one weight; presumably these tended to drag the float beneath the surface of the water. He himself had not experimented to see if the reel did in fact give a true vertical record, though he did suggest that it might be used with the ordinary lead whilst under way, that is, it could replace the waywiser for flying soundings. In practice it might have been difficult to judge the speed of a reel which was pitching on the waves at an ever-increasing distance from the mother-ship; at night it must have been impossible.

In 1806 Massey's second patent[22] listed various improvements to the sounder, chiefly dispensing with the bevel-rod and drop-piece to simplify the construction and to reduce friction (figure 27). In 1808 *Nicholson's Journal* reported that this sounder gave true readings in 30 fathoms from a vessel going at 6 knots, and that soundings could also be taken in 'very deep water' (here meaning in excess of 80 fathoms) where it was impossible to take them by the common lead.[23] Presumably, on such occasions the ship was hove-to.

In March 1811 the Committee of Mathematics and

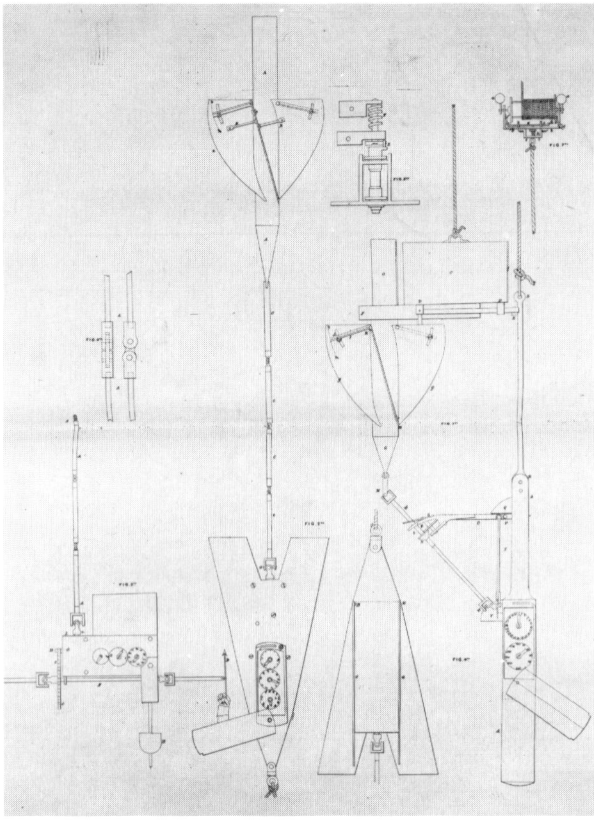

Figure 26. Patent drawing for Massey's log sounder, 1802 (*GB Patent Specification* No. 2601).

Mechanical Inventions at the Royal Institution reported on Massey's sounder,[24] and the chairman made plain his desire

> . . . to put on record the origin and progress of an invention which has been universally adopted by the British Navy . . . in hopes that this valuable invention may, through the extensive circulation of your magazine, meet with that publicity which it so highly deserves.[25]

He went on to say that, when sounding in water over 12 fathoms deep, it was difficult to heave-to in stormy weather when for lack of visibility the need for sounding was greatest. Two vessels tried out the sounder off the Isle of Wight in June 1806; other experiments in the English Channel had been under Massey's own supervision, while a series of testimonials from naval captains all confirmed that soundings of 60 fathoms were possible with the vessel proceeding at 6 or 7 knots. The Navy Board ordered 500 sounders in February 1807[26] and 250 some 18 months later. In May 1811 an order for another 1000 sounders made a total of 1750 for ships of the Royal Navy. (In 1814, at the close of the Napoleonic Wars, the Royal Navy had about 650 ships in commission.)[27]

It was discovered around this time that below 80 fathoms the air tubes of the rotators were frequently crushed by pressure. Massey tried perforating the tube to restore equilibrium between the interior and exterior, but this had the effect of introducing a registering error if the rate of descent was less than 17 fathoms per minute. Eventually he braced the air tube with brass discs, and satisfactory trials were made with the apparatus even when it was lowered to 200 fathoms. However, as late as 1838 damage to the air tubes at depths of 100 fathoms was still being reported,[28] so possibly these strengthened sounders were made only to special order.

In 1832 Massey went on board HMS *Trinculo* and sailed off the Irish coast to test three new types of sounding apparatus.[29] The first of these, a lineless detaching-weight sounder buoyed by a copper float, worked well in 100 fathoms, although its attached waywiser gave a higher reading than anticipated. Further tests made the following day, however,

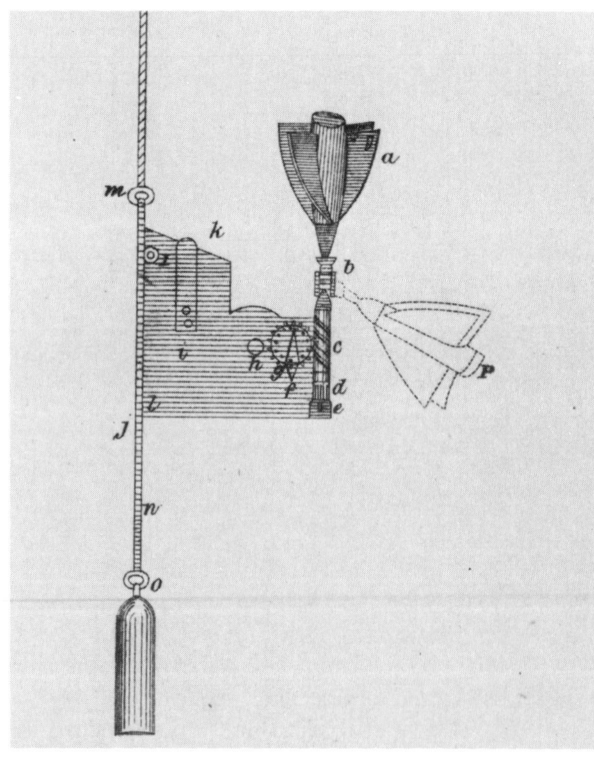

Figure 27. Patent drawing for Massey's sounder, 1806 (*GB Patent Specification* No. 2938).

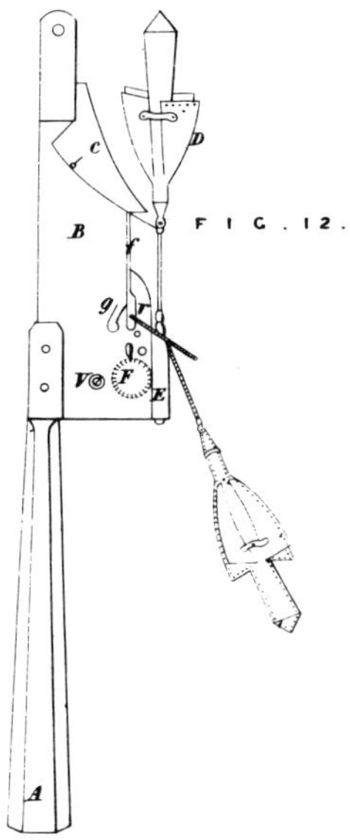

Figure 28. Patent drawing for Massey's sounder, 1834 (*GB Patent Specification* No. 6731).

resulted in the loss of four sets of apparatus; believing that their floats had been damaged by compression, Massey put a set of four spheres in place of the single one previously fitted, but with no happier result, for this apparatus too was lost.

He next tested a detaching-weight sounder to be sent down on a line. This apparatus seemed promising; its weights were lashed so that they broke free on hauling up, while the attached rope ensured that the sounder would not be lost even if its copper float was crushed. 540 fathoms of rope were veered overboard but when the sounder was recovered its waywiser had registered only 280 fathoms. Again Massey suspected that the problem was due to compression and he twice sent down copper spheres in a net, to 800 fathoms. The results were dramatic if inconclusive: in both cases the spheres vanished, leaving behind only their shredded container nets. The final apparatus tested was another lineless sounder of wooden construction; this too was lost.

The waywiser seemed the best method of measuring depth and in 1834 three new features were patented (figure 28).[30] Firstly, the register and figures were moved to a separate plate fixed to the side of the sounding lead. Secondly, there was added a stop-plate which fell sideways with the rotator and prevented it from turning when it was being hauled up. Thirdly, a swivel joint replaced the former universal joint so that the rotator now turned only during descent.

Inevitably, sounders as manufactured differed somewhat from patent drawings which concentrated on showing the design feature claimed as a new invention (figure 29). By 1838, however, Massey reported a version that worked well under way in depths of 80 or 90 fathoms.[31] The base of its lead was armed with tallow to confirm that bottom had been reached and to indicate the nature of the sea bed. In 100 fathoms or

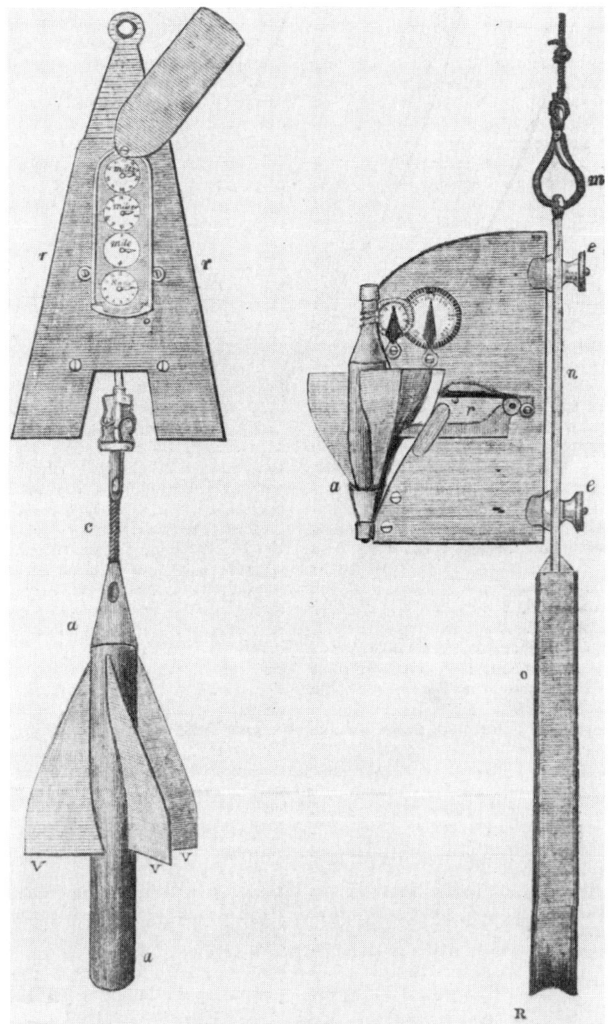

Figure 29. 'Massey and Windham's perpetual log and deep-sea lead'. In Fig. 2, o is the lead, extended by plate n. a is the rotator, r is the locking plate which drops when the sounder is hauled up, and ee are screws fastening the sounder and its plate to the lead shaft. From Whitley Baker 1838 *Mech. Mag.* **28** p241, Figs 1 and 2.

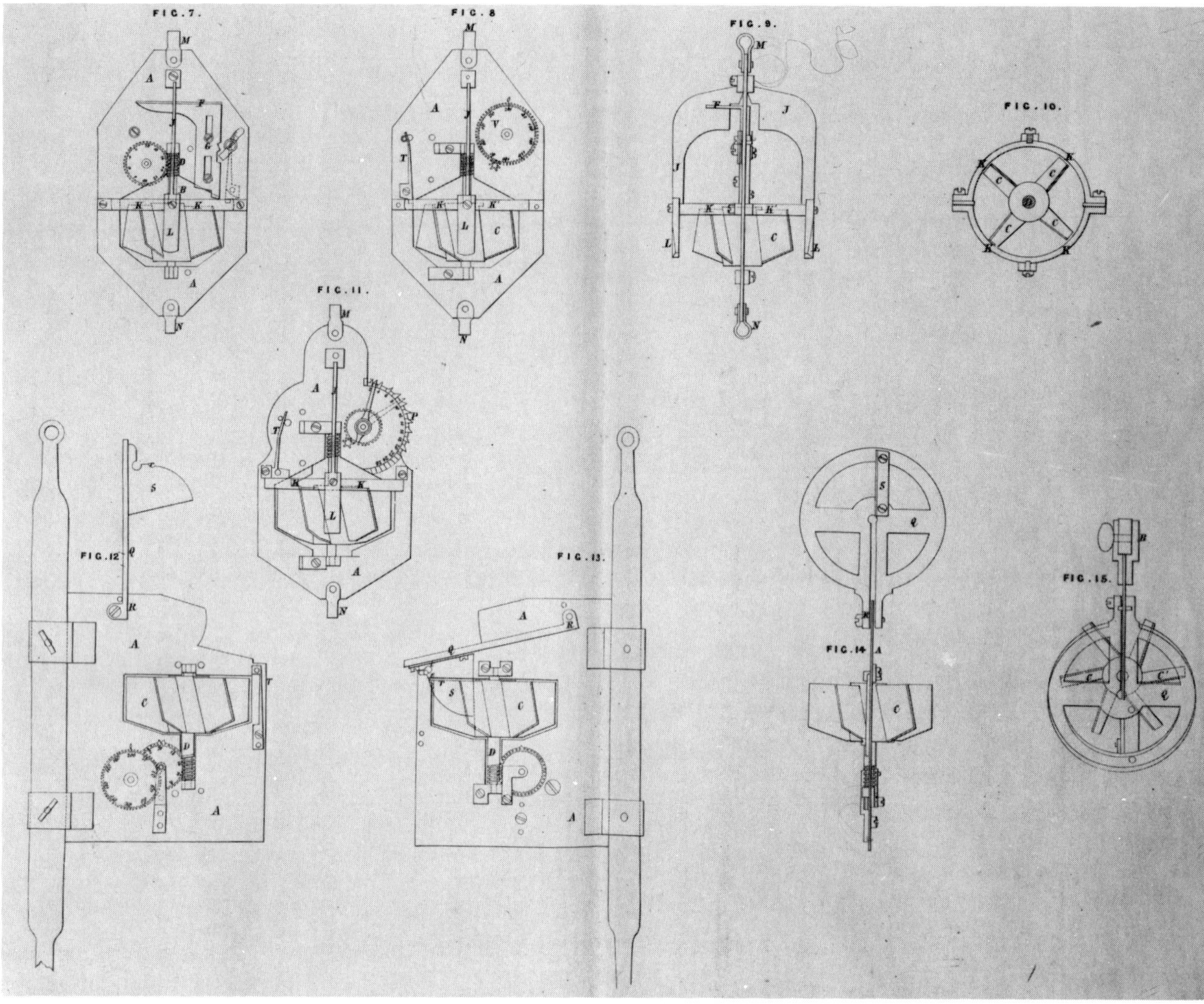

Figure 30. Patent drawing for Massey's sounder, 1836 (*GB Patent Specification* No. 7113).

more the rotator was apt to burst and, while this was immaterial for navigators working over shallow water, it was a fault which Massey had to overcome before he could construct a true deep-water sounder.

The final patent of this series, dated 1836,[32] featured a sounder carried on a line vertically above and separate from the lead (figure 30); a stop to prevent the vanes from turning before the sounder entered the water, and a version in which part of the frame was cut away so that the dial might be 'read' by touch alone, at night for instance. The patent drawing also showed some of these details adapted to the conventional sounder affixed to the side of the lead.

3.4. BURT'S BUOY AND NIPPER

At this point it is timely to introduce Peter Burt's buoy and nipper which was an aid to sounding operations, and at the same time to look at the term 'sounding-machine' which was used with a variety of meanings, thus giving rise to some misunderstanding among historians ever since the nineteenth century.

When sounding by simple lead and line it is essential that the gear should run vertically down from ship to sea bed since the length of line in the water is taken as the true measurement of depth. In shallow water, with the ship under way, the leadsman obtained a vertical cast by heaving the lead from the bows and walking back to the stern as the ship went forward, keeping pace with the point at which the lead had entered the water. If the sea was more than 10 to 20 fathoms deep, however, and the vessel was small or moving fast, this method naturally became impossible. Compensation might be made for the angle of the line, but this was difficult to observe at night or in bad weather—those very times when safe navigation depended on accurate soundings.

Massey's 1802 patent incorporated a floating frame to keep the sounder over the point of entry, and a similar patent was filed by Robert Dickinson in 1814,[33] but neither of these would have been easy to use at night. Burt's was a far simpler device and, although sometimes referred to as his 'sounding machine' and contrasted with Massey's 'sounding machine', it consisted only of an inflatable canvas bag or buoy about two feet long, to which was attached a wooden arm with a spring-loaded sheave and nipper (figure 31). A common tagged lead-line was rove through the pulley and the whole apparatus cast overboard, the line being paid out as the ship ran on, leaving the buoy over the point of entry. As the lead reached bottom and the strain came off the line, the nipper clamped against the sheave and the entire apparatus could then be hauled in. The amount of line beyond the nipper thus represented the vertical distance between sea surface and bottom.

Burt's apparatus seems to have been introduced shortly before 1818, when it was tested by the Admiralty against Massey's sounder for use by ships under way. Much controversy was generated;[34] articles and pamphlets were published by both parties and their supporters mainly arguing over the value of Burt's device to navigation. However, it did come to play a vital part in deep-sea sounding operations and was issued on many expeditions during the first half of the nineteenth century.

The term 'sounding machine' which was applied indiscriminately to many of the devices from the time of Hooke onwards, was also used for Burt's buoy and for the Massey waywisers. In the 1870s the term was restricted to the machinery of a steam-powered reel incorporating a spring, from which wire line plus detaching sinker was automatically paid out and braked when the bottom was reached. The earlier use of the term to mean any device, whether mechanical or pressure-operated, was dropped. The difference between Massey's and Burt's apparatus was therefore more properly one of method rather than of machine since it was possible, indeed beneficial, to employ the two together when sounding.

Figure 31. Burt's 'buoy and nipper'.

REFERENCES AND NOTES

1. Cavendish C 1770 'A description of some thermometers for particular uses' *Phil. Trans.* **50** pp300–10
2. de Saussure H-B 1796 *Voyages dans les Alpes* (3 vols) vol. 3 pp196–201
3. de Saussure H-B 1796 *Op. cit.* p196
4. Six J 1783 'An account of an improved thermometer' *Phil. Trans.* **72** pp72–81; also 1784 'Experiments to investigate the variation of local heat' *Phil. Trans.* **74** pp428–36
5. Six J 1794 *On the Construction and Use of a Thermometer* . . . (Maidstone)
6. For a more detailed account of Six's life and scientific writings see Austin J and McConnell A 1980 'James Six, FRS: two hundred years of the Six self-registering thermometer' *Notes and Records of the Royal Society* **35** (1) pp49–65

7. Liebnitz G 1745 *Vivorum Celeberr. Got. Gul. Leibnitii et Johan Bernoulli Comm. Phil. et Math.* (Geneva) vol. 1 pp371–8
8. Six J 1794 *On the Construction and Use of a Thermometer* . . . (Maidstone) Ch. 8
9. Surviving experimental forms of this thermometer are described in (5)
10. Aboville A-M 1798 'Changement proposé au thermomètre de l'Anglais Six pour le rendre propre à connaître la température des eaux à des grandes profondeurs' *J. Mines (Paris)* **9** pp75–6
11. von Krusenstern A 1813 *Voyage Around the World* . . . (London) Introduction pp8–9
12. von Krusenstern A 1813 *Op. cit.* pp202–3
13. Prestwich J 1874 'Tables of temperatures of the sea at different depths beneath the surface, reduced and collated from the various observations made between the years 1749 and 1868 discussed' *Phil. Trans.* **165** pp587–682
14. Arago F 1840 'Températures sous-marins' *C. R. Acad. Sci. Paris* **11** pp310–15
15. Charnock J 1795 'Description of a sea gage for the purpose of sounding in currents and great depths of water' *Rep. Arts and Industry* **2** pp180–4
16. de Saumarez H 1725 'Account of a new machine called the "marine surveyor"' *Phil. Trans.* **23** pp411–14
17. *GB Patent Specification* No. 1028 (1772)
18. *GB Patent Specification* No. 1895 (1792)
19. *GB Patent Specification* Nos 2405 and 2458 (1800)
20. Treherne A 1977 *The Massey Family, Clock, Watch, Chronometer and Nautical Instrument-makers* (Catalogue of an exhibition at the Museum, Newcastle-under-Lyme, Staffs)
21. *GB Patent Specification* No. 2601 (1802)
22. *GB Patent Specification* No. 2938 (1806)
23. Anon 1808 *Nicholson's Journal* **21** pp242–55
24. Massey had already received premiums from the Society of Arts, Manufactures and Commerce for improvements to clocks
25. Clifford R 1811 'Report on the new sounding machine invented by Ed. Massey made to the Royal Institution, 4 March 1811' *Phil. Mag.* **40** pp91–9
26. See (23), but Captain R J Neve, writing from HMS *San Joseph* testified that the quality of the line issued by the Admiralty was not good enough for deep soundings
27. Archibald E 1968 *The Wooden Fighting Ship* (London) Appendix 4, p154
28. Whitley Baker E 1838 'Massey & Windham's patent perpetual log and deep-sea lead' *Mech. Mag.* No. 753 p244
29. Originally an article in the *Cork Reporter,* it was reprinted in *Naut. Mag.* **1** (1832) pp498–9
30. *GB Patent Specification* No. 6472 (1834), attributed to Edward Massey Senior. His father had died in 1813 and his son (born in 1799) was now in the same trade.
31. Anon 1834 *Repertory of Patent Inventions* **4** pp131–8; see also (28)
32. *GB Patent Specification* No. 7113 (1836)
33. *GB Patent Specification* No. 2866 (1814)
34. See for example Anon. 1819 *Q. J. Sci., Lit., Arts* **6** p135; Burt P 1819 *Copies of Reports of Experiments made . . . for the Purpose of Ascertaining the Superiority of Burt's Sounding Buoy and Knipper over Massey's Sounding Machine* (London) and Massey E 1820 *A statement of the case of Mr Edward Massey . . . most respectfully offered to the notice of every member of Parliament.* (Prescot)

Chapter 4

The Arctic Expeditions, 1773–1828

4.1. NAVAL OCEANOGRAPHY ON PHIPP'S NORTHERN VOYAGE (1773)

Ever since the sixteenth century shipmasters had hoped to discover an Arctic seaway linking the Atlantic and Pacific Oceans. Trading voyages into the Pacific revealed potential markets rich in spices, precious metals and furs. The journey from Europe via the Magellan Strait or past the Cape of Good Hope to the Indies was lengthy and it was therefore worth seeking a northern route, although the northern part of America itself did not at first attract traders.

By the middle of the seventeenth century it was clear that neither the Gulf of St Lawrence nor Hudson Bay opened into the northern seaway, but if one was merely the mouth of a great river and the other a huge but apparently closed bay, both gave access to the great forest belt of northern America. Within this forest lived beaver, bear, ermine, mink, lynx, and the Indians who were prepared to trap them and sell the skins to European traders.

This forest belt, which now forms part of Canada, was first penetrated from the south by French explorers. The English controlled most of the north, mainly through the agency of the Hudson Bay Company, which was set up by Royal Charter in 1670. The chartering of companies by the Crown had been an established method of encouraging trade and territorial expansion for the past century and, in exchange for sole rights in trade and commerce, the company was given a great sweep of territory surrounding Hudson Bay itself. The forested zone, which alone yielded the furs which were the original reason for the foundation of the company, did not in fact extend north beyond the southern shores of the Bay. The company was expected, and did intend, to find an Arctic seaway which would allow them to expand their trade routes.

Early hopes rested on finding the passage within the confines of Hudson Bay, north of the company's original settlements, but as time progressed, profiting from the developing trade in copper, furs and other local products, the company seems to have had ambivalent views on this project. Although they were keen to extend their own hegemony, they were reluctant to publicise their discoveries lest other ship-owners gain advantage from their findings. Uncertain which parts had or had not been explored, the Admiralty sent Captain Middleton with HM Ships *Furnace* and *Discovery* into the Bay in the autumn of 1741. On his return late the following year Middleton claimed that, though tidal observations led him to believe that he had reached the entrance of a strait, it had been so encumbered with ice as to be unpenetrable. Put briefly, the sequel was that Middleton's story was not believed, and it was suggested that the Hudson Bay Company had bribed him to conceal his discoveries and the Admiralty court found him culpable of this charge. The following year Parliament passed an Act offering a reward of £20 000 to any of His Majesty's subjects who should discover a northwest passage through Hudson Strait to 'the western and northern ocean of America . . . to the benefit and advantage of the trade of this Kindgom'.

Tempted by this offer, Captains Moor and Smith took *Dobbs* and *California* to find a northwest passage in 1746. They were accompanied by Henry Ellis, whose experiments with Hales' bucket were related above (see §2.4). After wintering within Hudson Bay, the tides and currents were again observed and the ships entered

Wager Strait, where they sounded to find no bottom with 140 fathoms of line. They sank an empty bottle, bringing it up full of water 'as salt as the Atlantic'. However, they soon found shoal water and the inlet was terminated by two unnavigable rivers. With the season drawing to a close, the expedition returned home in the autumn of 1747.

During the eighteenth century the Admiralty despatched several pairs of ships on voyages of exploration into other distant waters. Frequently their crews had some specific task to fulfil, designed to improve the art of navigation. Perhaps the most famous of these voyages was that of Captain Cook who, in 1769, sailed with a single ship, HMS *Endeavour*, into the Pacific Ocean where he was to observe the transit of Venus from the island of Tahiti. Cook's orders were to explore the southern oceans and carry out the astronomical observations, which kept him to a strict timetable. He did make some measurements of the upper sea temperature and his naturalists lowered a thermometer on a few occasions, but he was kept busy making charts and seems understandably not to have taken an interest in the depths of the ocean.

In the same period, the Royal Society encouraged the Admiralty to send a naval expedition to search for a route via the Arctic to the Pacific Ocean. It was believed by some persons that open water lay up to the Pole (such rumours were apparently circulated by Dutch and other foreign sailors), in which case a very direct track to the Bering Strait could be followed. The Admiralty prepared two sturdy vessels of the type known as 'bombs': HMS *Racehorse* with a crew of 90 men was commanded by Constantine Phipps (later Lord Mulgrave), and the smaller HMS *Carcass* (this curious name refers to a type of bomb then in use) sailed from the Nore in June 1773 (figure 32).[1]

Phipps took with him various geophysical instruments in the hope of resolving certain problems of navigating in high latitudes where the proximity of the Earth's magnetic pole made compass readings unrealiable. In the event, the voyage lasted only a few months because he was unable to penetrate the ice-pack

Figure 32. HM Ships *Racehorse* and *Carcass*.

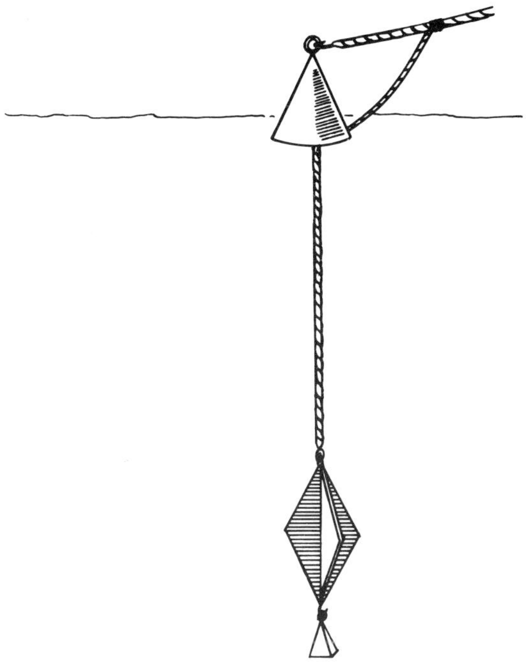

Figure 33. Bouguer's log adapted for current measurements. Redrawn from Bouguer 1747 *Mem. Acad. Royale des Sciences* (Paris) plate 20.

between Svalbard (Spitzbergen) and eastern Greenland, and the onset of bad weather in August forced him to turn for home. On passage out and back, he took advantage of calm weather to investigate the depths of the sea and for this had been issued with several instruments, among them a Cavendish minimum thermometer, Bouguer's log (which doubled as a current meter), and a waterbottle of unspecified type. He also received from Cavendish's son a set of correction tables for the thermometer to offset the distortion of the alcohol by pressure. Phipps described one of the operations in his narrative:

> June 20th: Almost calm all day. The water being perfectly smooth I took this opportunity of trying to get soundings at much greater depths than I believe had ever been tried before. I sounded with a very heavy lead the depth of 780 fathoms, without getting ground; and by a thermometer invented by Lord Charles Cavendish for this purpose, found the temperature of the water at that depth to be 26° of Fahrenheit's thermometer; the temperature of the air being 48½°.[2]

On the preceding day Phipps had found his position to be just south of the Arctic Circle and slightly west of the Greenwich meridian. In this area the depth of water exceeds 1000 fathoms. Nine days later he again sounded and also tried the current:

> June 29th: At three in the afternoon, brought to and sounded 110 fathoms; soft muddy ground: hoisted out the boat and tried the stream, found it, both by the common and Bouguer's log (which agreed exactly) to run half a kt N.[3]

Bouguer's log had been described in 1747;[4] it could be adapted to measure currents under the surface by hanging a vaned 'diver' below to be driven by the sub-surface stream, relative to the observing boat (figure 33).

On another occasion Phipps sent down a Cavendish thermometer to 115 fathoms, accompanied by the waterbottle. The thermometer registered 33° but the waterbottle sample, measured on deck, was 38½°, far closer to the surface temperature of 40° and the air temperature of 44½°. Aware that the pressure correction tables he was using might be inexact, Phipps then tried an alternative means of testing the temperature.[5] Assisted by Dr Charles Irving, the ship's surgeon, Phipps insulated a bottle by wrapping it in jackets of wool, oilskin, leather and tarred canvas (figure 34), plugged with a small lead cone hung with strips of leather, but even then, Phipps had no means of telling which of his temperature measurements was closer to reality. With hindsight we now know that his thermometer was unreliable at depth; that in the Arctic Ocean the deepest water is not always the coldest, so that minimum thermometers are unsuited to measure it; and that Irving's waterbottle was unlikely to have recovered water from exactly the depth intended.

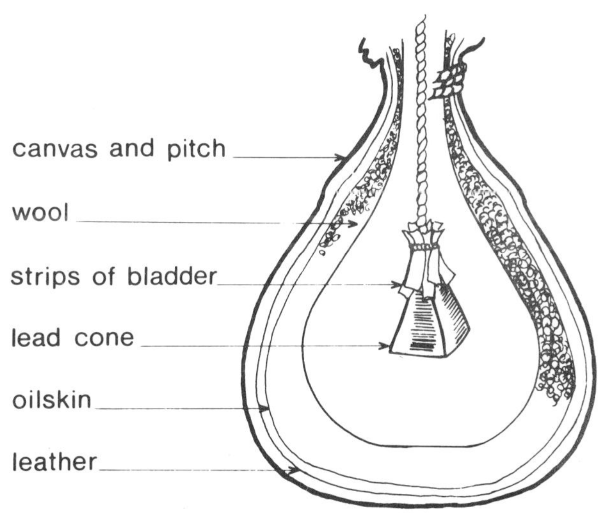

Figure 34. Phipps' insulated water bottle (tentative reconstruction).

4.2. 'PRIVATE RESEARCH': WILLIAM SCORESBY (1810–22)

After Phipps returned to England in 1773 the Navy took little further interest in Arctic exploration. The period around the turn of the century was marked by wars in which most European nations participated, but hydrographers were still sent to chart areas where warships engaged in patrols, blockades or battles. Commerce continued, however, and each summer British ships sailed to the Greenland whale fishery. Two such ships, the *Esk* and the *Baffin,* belonged to William Scoresby of Whitby whose son, William junior (1789–1857), sailed each summer in company with his father. Scoresby senior was himself of a scientific bent and an acquaintance of Sir Joseph Banks, then President of the Royal Society. Young William was sent to attend lectures at Edinburgh University each winter where he studied natural philosophy under Professor Jameson. Each summer the voyages continued, so that theoretical and practical learning combined to give Scoresby an unrivalled knowledge of the Arctic. His interests ranged over its weather; the effects of sea temperature and currents on the distribution of the whales; geomagnetism (the proximity of the magnetic pole made compass navigation hazardous), and the charting of the coasts along which he sailed. He belonged to various learned societies and published his findings in their journals as well as conversing with eminent scientists of the day. Before he abandoned the seafaring life to enter the Church he wrote two books: the results of his voyages beginning in 1810 were published as *An Account of the Arctic Regions* (1820), while a subsequent voyage was described in *Journal of a Voyage to the Northern Whale Fishery* (1822).[6]

Scoresby's career is illuminated for us because he bequeathed his papers and some apparatus to the Literary and Philosophical Society of his former home town of Whitby. From his university notebook we can learn details of the instruments and methods that were in use during the period between Phipps' voyage and those of the Admiralty-directed Arctic explorations which began in 1818. His informal correspondence also amplifies the dry accounts of his experiments as related in his books and articles.[7]

Professor Robert Jameson, under whom Scoresby studied, had himself been a pupil of John Walker whose lecture topics in natural history given between 1779 and 1803 have survived.[8] Walker's discourse on hydrography was brief, and that on the waters of the sea merely formed part of a wider survey encompassing rivers, lakes and subterranean water. Walker was not in a position to give many facts for, as he said, 'There is no bottom to be discovered in the ocean at any great distance from land.'[9]

Within the West Indies, between the Guinea Coast of Africa and the American shores, ships were out of soundings soon after leaving land. Walker did report on temperatures, citing Ellis's use of a Hales bucket and the discovery that temperature fell gradually to 53 °F at 3900 feet, remaining constant below that level. In another series of lectures Walker suggested that the depth of the sea might correspond to the heights of mountains and he declared that no plants were found in deep water apart from the Sargasso weed, nor did fish live below the upper levels except in the 'torrid zone'.[9]

Scoresby's lecture notes reveal that by the time he got to Edinburgh some advance had been made in temperature measurement, and the insulated cladding that Phipps and Irving had devised was now being applied to other types of sampler:

> Lecture given 18 December 1809
> *Temperature of the sea at different depths*
> For this purpose different instruments have been contrived but generally agreed in one point that is bringing water from great depths and so ordered not to allow the temperature to change in any moderate time, this is effected by having the vessel containing the water covered with wool then leather then perhaps some other substance ill conductor of caloric and the whole contained in one or more pitched canvas bags. The method used to get the water is by having valves in the apparatus so that when it descends the water has a free passage through it but when it ascends the valves are closed. Another apparatus has been constructed for securing the temperature from changing by having a glass tube enclosed in a wooden one twice its dimensions and the intermediate space filled with powdered charcoal, this was put into another larger case and the space between filled with tallow; this is enclosed in a waterproof leather case and the whole in a pitched bag.[10]

Other notes cite figures for sea temperature extracted from Phipps' results and from the experiments of de Saussure using his insulated thermometer (see §3.1). The business of sounding was soon dealt with:

> The instruments that are used for this purpose are the deep-sea line and the sea-gauge. The bottom of the sea like the surface of the land is marked by irregularities.

For his own early experiments at sea, Scoresby used a Hales bucket, adapted from a ten-gallon cask made of fir plank, to which he added an external flap to help keep the valves closed as it was raised. A common thermometer served to test the temperature when it was back on deck. After several excursions, however, the wood became waterlogged and the staves split. Scoresby therefore approached Banks to provide something better and the latter passed the request on to Henry Cavendish, the same that had prepared the compensation tables for Phipps' thermometer. Unfortunately Cavendish died before the work was completed, but in a letter to Scoresby dated 8 September 1810, Banks said that the instrument-maker Cary was constructing the apparatus and that it would soon be ready for collection. Banks also gave a Six thermometer which was fitted into the sampler, the first account of this instrument being used on a British ship. However, this sampler fared no better, as Scoresby related in a letter to his old teacher Jameson in September 1811:

> . . . Sir J Banks agreeable to his promise furnished me with a neat apparatus made by Carey[11] under the instruction of Cavendish and others for the purpose of ascertaining the temperature of the sea at considerable depths in Greenland, connected with it goes a thermometer for showing the range of temperature in descent. The thermometer acted its part admirably but the rest of the apparatus tho' made at some expense was not sufficient to withstand the immense pressure of the waters at such depths as 600 or 800 feet. The deal of which it chiefly consisted was completely soaked with water and tho' bound with immense brass hoops the seams were opened and the valves rendered leaky. The results obtained with it were the same as those I sent you last year. At the depth of 100 fathoms the water being 5–6° warmer than at the surface.[12]

Scoresby must have considered the design an improvement, however, for the story of its development continues:

> I made a model of a similar instrument and got it cast in brass. This I fitted up, and applied to it the valves made by Carey, which then proved an elegant and useful apparatus.

Scoresby's 'marine diver' was smaller than his previous samplers and beside the sprung valves, it had two plate-glass panels so that the Six thermometer could be read before the container was opened. It weighed 23 lb and so did not require extra sinkers to hasten its descent. By fixing wire gauze across the upper opening Scoresby was able to capture small fish and other organisms that were swept into the sampler, as a bonus for his researches (figure 35).

From time to time harpooned whales sounded to the bottom, carrying harpoon and line with them; on one such occasion the whale took with it the ship's boat. After much effort the whale and boat were brought to the surface and the boat was found to be so waterlogged as to be unusable even as fuel, being rejected as incombustable by the ship's cook. This experience led Scoresby to carry out experiments with different types of wood and other substances, lowering them in his marine diver and comparing their weights afterwards with samples of identical materials immersed in a bucket of water on deck for the same period. He was engaged on one of these trials, which he believed to represent a new and promising line of enquiry when, on 28 June 1817, the marine diver was lost while it was being hauled up from the greatest depth yet sampled.

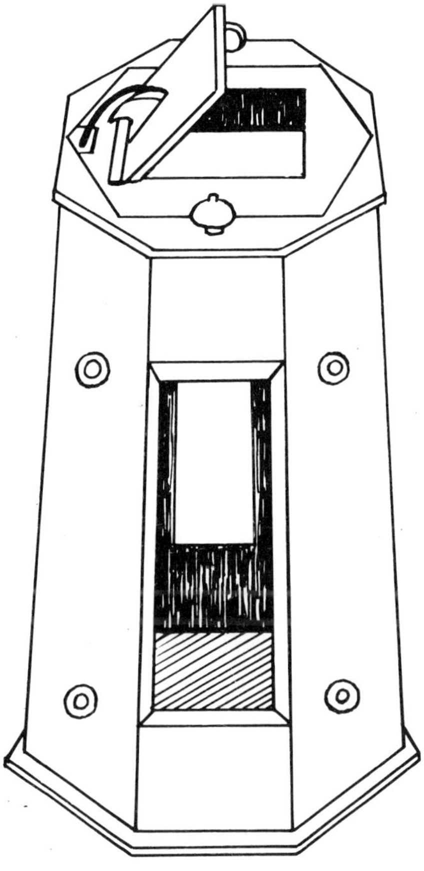

Figure 35. Scoresby's 'marine diver'. From Scoresby 1820 *Arctic Regions* vol. 2, plate 2, Fig. 2.

Behind the laconic report in his book lies the story of a complex experiment which went wrong due to a very small mishap. His great disappointment was revealed in a letter to Jameson written soon after the event. Here is Scoresby, still rather breathless, recounting the day's events:

> . . . I wished to try the effect of pressure at the depth of a mile or a mile and a half. I therefore prepared for the experiments by providing blocks of wood of different shapes . . . and other substances to the amount of 20 articles . . . I had other matters also in view—the temperature of the sea at a depth before scarcely sounded was to be ascertained. The water of greatest depth to which the apparatus was sent was to be brought up, its specific gravity and constituents examined—two tin vessels sent by Mr Adie for trying the depth and mean specific gravity of the water passed through in the descent to be proved. The nature of the current to be examined and by means of a frame of wire gauze stretched across the upper valve of the marine diver it was converted into a trap for small fishes . . . I procured all the lead lines I could meet with and having a favourable opportunity on 28th June I moored the ship to a field of ice. Fixed the whole apparatus consisting of the marine diver, a Six's thermometer, Mr Adie's tin vessels, the specimens of wood, bones, jet, metals etc and allowed them to sink to the perpendicular depth of 7200 feet or upward without meeting the bottom. (The lines being carried obliquely by the current, I fastened it to a boat and committing it to a judicious officer, sent him to find the perpendicular of the lines. This was accomplished and the ship worked towards the place and the end of the line after two hours resting taken with some trouble to the ships, for on account of the current it proceeded to windward as fast as the ship could work. By means of a whale line the coil of line resting in a boat was after an interval of two hours taken aboard the ship whereby a slow and steady motion we got hold of the boat, hauled it to the ship, secured the end of the line and proceeded by a slow and steady motion began to haul it aboard.) We had drawn in about 300 yards when to my extreme mortification the line gave way, slipped through the grasp of the man who held it at the time, with the velocity of lightning and disappeared in a moment! Thus . . . an experiment in which I have placed such sanguine expectations was blasted. The apparatus set a high value up being the only one of its kind in existence . . . It was very well finished and a beautiful instrument. It was fitted with the Six thermometer and the valves of the original instrument. I cannot say whether the loss of the experiment or the loss of the instrument gave me most concern. The apparatus and 1050 fathoms of line which accompanied it were of the value of £20. The line which broke was the thickest and apparently the strongest of the whole lines in use. A small part scarcely 2 inches in length proved to be injured by moisture and was *rotten*. The weight certainly exceeded by much what I had calculated. The rope being thoroughly wetted became near half as heavy in water as it was before in air. Thus terminates my experiments on the temperature of the sea. I suppose the apparatus has found the bottom ere now![13]

Poor Scoresby; according to his log book the marine diver rests on the bottom in lat 76°30′ N, long 4°48′ W. He did continue his experiments, though he never had another diver made, relying on the Six thermometer alone on his later voyages.

Scoresby recognised that each layer of the sea contained its own currents. He took evidence from the surface drift which floated objects from known sources to distant shores; from the tracks of icebergs propelled by shallow sub-surface currents acting on their submerged bulk; and from the deep-water movements which in places up-welled as perceptible warmer water-masses. As did the whales themselves, Scoresby distinguished the various waters by their temperature, salinity, colour and planktonic life.

When sounding, accepting that it was very hard to tell when a heavy lead and line had reached bottom, Scoresby used a light assembly consisting of a 20-lb sinker and a 'very small line' so that the lead was one-third of the total weight and therefore the relief on the line as the sinker grounded was easily felt. With this gear he claimed to have sounded in excess of 1000 fathoms. However, despite his belief in the existence of submarine currents he sometimes sounded with a far greater amount of line than the true depth; indeed, on the aforementioned sounding of 1000 fathoms he failed to reach bottom. Many years later Scoresby gave a talk at a meeting of the British Association for the Advancement of Science of his experiences in sounding and the way in which a line could be swept far out of plumb by submarine currents.[14]

While still experimenting with the effects of compression at depth he had

> hopes that the degree of impregnation . . . might be applicable as a measure of depth. If this were the case, it would serve a very valuable purpose, since all the plans hitherto contrived for measuring depths from a vessel, when sailing slowly or drifting through the water, cease to be useful beyond 200 or 300 fathoms.[15]

In a footnote he added:

> A very recent contrivance, for obtaining soundings from

a vessel under way, founded on the small change in bulk which takes place in water when strongly compressed, promises to answer the design at still greater depths than 200 or 300 fathoms.

This referred to a sounder proposed by Jacob Perkins which, though tested, was not developed further.[16]

4.3. JOHN ROSS AND THE 1818 ARCTIC EXPEDITION

The Napoleonic Wars came to an end in 1814 when the victorious nations met in Vienna to redraw interstate boundaries in the Old World, but far-reaching events were also taking place in the New World, where Spain and France had been too preoccupied with home affairs to prevent the break-up of their American colonies: the French island of Haiti in the Caribbean and the Spanish viceroyalty of La Plata were the first to achieve independence, but the rest of Central and South America was already in a ferment. To the north, the Pacific shores were under Spanish control as far as California, while Russian settlers were moving down from their Alaskan sealing grounds. A new wave of migration carried the western frontier of the United States across the Rockies and down to the Pacific Ocean, establishing a wedge between the other groups, and gradually displacing them.

As the Spanish trading monopoly in the Americas withered away, new markets could be opened up by British merchant ships, and Royal Navy vessels arrived to survey coasts and harbours around the continent. Inevitably, attention was once more directed towards the elusive northwest passage, especially as Russia was now probing for it from the Pacific end. Success appeared to be within reach, for there was some climatic improvement within the Arctic which caused large masses of ice to break away from the main pack and drift southwards. This amelioration was noted by whaling captains and Scoresby wrote to Banks about it, reporting that for the first time for many years the eastern coast of Greenland between 74° and 80° N was clear of ice. The opportunity was not to be wasted. A secretary at the Admiralty, John Barrow, who as a young man had sailed in Arctic waters, persuaded the First Lord (Lord Melville) that the time had come to send out another expedition. To stimulate discovery, Parliament offered a reward of £5000 to the crew of any ship making 100° west of Greenwich.

In the spring of 1818, therefore, a fleet of four ships was despatched northwards in another attempt on the passage. After proceeding together as far as Davis Strait, HM Ships *Dorothea* and *Trent*, under the command of David Buchan and John Franklin, respectively, were to keep a northerly course through the Spitzbergen seas as far as they were able. The remaining pair, HMS *Isabella* under the expedition's commander Captain John Ross, and HMS *Alexander* under Lieutenant W E Parry, were to turn into Davis Strait and from there were to seek a route north of the American continent through to the Pacific Ocean.

Ross's instructions described the oceanographic effects to be observed should such a passage be suspected and in this way indicated the marine investigations that he should undertake.[17]

From the best information . . . a current of some force runs from the northward towards the upper part of Davis' Strait . . . This current, if it be considerable, can scarcely be altogether supplied by streams from the land, or the melting of ice, there would therefore be reason to suppose, that it may be derived from an open sea; in which case Baffin's Bay cannot be bounded by land, as our charts generally represent but must communicate with the Arctic Ocean.

In passing up the Strait, if such a current should be discovered, it will be of the greatest importance to you . . . leading you direct to the opening by which it may be supposed to pass from the Arctic Sea into Davis' Strait.

The strength and direction of the current should be tried once in 24 hours; or oftener if any material change is observed to take place; and it will be most advisable to take its temperature at the surface frequently, as you proceed, to compare it with the temperature of the surface, where there is no current.

Other tasks were detailed, for the improvement of geography, navigation, hydrography and science generally. The best available instruments were provided and Captain Edward Sabine of the Royal Artillery was seconded to *Isabella* to assist with the observations. Ross was ordered to measure temperatures at different depths, to note the velocity and set of all currents, both tidal and constant, and to sound and collect bottom samples. For this he was issued with:

Ten thermometers, Fahrenheit, by Jones.
One self-registering thermometer, Fahrenheit, by Jones.
Apparatus for taking up seawater from given depths above 80 fathoms.
Bains' patent log.
Massey's patent log.
Jennings' patent log and glass.
Burt's buoy and nipper.
Deep-sea leads: one each of 150 lbs; 100 lbs and 50 lbs weight.
Copper floats in which cards giving the date and position of the vessel were sealed before the floats were cast overboard.

The omission of Massey's sounder is surprising in view of the Navy's considerable purchases of this apparatus.

The 'apparatus for taking up seawater' had been specially designed by Sir Humphry Davy for detecting the presence of a strong current of saline water, less cold than that at the surface, indicative of a channel leading to a polar sea.[18] Made by the instrument-maker Newman, it was a simple, pear-shaped copper container some ten inches long, provided with a valve at the top and a plug at the base which could be unscrewed to drain the contents of the bottle when it was brought back on deck (figure 36). The opening and closing mechanism, which was operated by pressure, was complex and depended on the action of a piston operating a series of levers. The piston moved within a cylinder cut by notches to represent pressures of one, two, three or more atmospheres. When the bottle was cast into the sea it sank with the valve closed, until the pressure equalled that for which it had been set, when the valve would open. While still descending the bottle filled with water, and the piston completed its travel and closed the valve; the apparatus could then be hauled up. The bottle's operating range was between 5 and 80 fathoms.

Ross did not give much detail of the bottle's use but in an appendix to his narrative he noted:

> *Sir Humphry Davy's waterbottle*
> Did not close so as to prevent the water from escaping or mixing with that nearer the surface as it came up.[19]

The bottle was not insulated and temperature measurements were made by lowering the Six's thermometer on the sounding line. Many years later Sabine recalled its use when he discussed with Prestwich how they had recorded the exceptionally low temperature of 25.75 °F.[20] On this occasion the thermometer had been recovered from a depth believed to be 4080 feet and its tin case had been full of water. This low temperature may have been that of the surface water since, if water had been forced into the case, the bottom temperature would have been raised under compression to a higher figure.

Each ship carried only one self-registering thermometer, but that on board the *Isabella* was broken by September. To continue his observations Ross presumably commandeered the thermometer issued to the *Alexander*. Ross's estimates of depth are often suspect. However, later in the course of the voyage he made a habit of putting a thermometer into the mass of sediment retrieved from the sea bed and found that these readings agreed 'tolerably well' with those of the self-registering thermometer. It seems likely therefore that the tin case usually protected the thermometer from compression, although this was apparently not deliberate.

The Admiralty had given Ross a sounder–sampler 'better calculated to bring up substances than the lead usually employed for this purpose', but the apparatus, designed by John McCulloch, ship's surgeon, failed to live up to their expectations.[21] Certainly Ross made

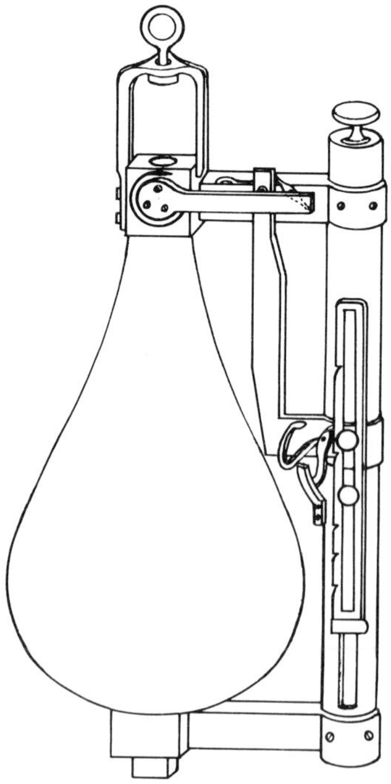

Figure 36. Davy's waterbottle, 1818. Redrawn from Davy 1818 *J. Sci. Arts.* **5**.

many fruitless attempts to sound and recover bottom samples, and by July had himself designed and made a model of a grab which his armourer constructed in the ship's forge (figure 37).[22] Ross termed his new grab a 'deep-sea clamm' (figure 38, *right*). It had a cast-iron outer casing 18 inches long, containing a pair of jaws which hung open during descent and were driven by the clamm's weight some distance into the sediment. On hauling up, the jaws closed and retracted the sample within the casing, in which position it was brought to the surface. The first one made weighed 1 cwt but Ross considered that in shallow water such as the North Sea, ½ cwt would suffice to drive the grab into the sediment. Describing its use, he wrote:

> . . . it is necessary to be provided with whale lines, such as are used by the Greenland and South Sea ships, which are of 2½ inches circumference, made of the best hemp and very pliable and easily coiled so as to run quite clear on the forepart of the ship's decks.
>
> In very deep water it is necessary that it should be calm or nearly so, to be certain that soundings are obtained in 500 fathoms, but in a light breeze may be hung to in a boat and towed in the direction of the ship's drift, and if there is any wind it is best to lower all the sails down. An outrigger, fitted with a block, should be fixed on the weather quarter, through which the line ought to be rove and bent to the instrument, when it ought to be lowered until it is a fathom below the surface and then let go . . .[23]

The waterbottle which Ross also devised to replace the unsatisfactory Davy bottle, was not constructed until his return. This 'hydraphorus' (figure 38, *left*) was as complex in design as that of Davy, but is of interest since it was the first of its kind to make use of a vane or propeller to close the valve. It had a cylindrical copper body, with a small opening to admit water, covered by a circular plate with a toothed rim and a hole of the same size. The plate was rotated by the vane and could be preset to bring both apertures into alignment at a selected depth so that water entered, expelling air through a spring valve at the top of the bottle. The same valve could be opened by hand on recovery, and the water was drained off via a stopcock.[24]

In deeper water additional weights could be added, attached to the lower rings which were for that purpose. Care was needed to get the waterbottle into the sea without damaging the vanes, and it was then lowered beyond the selected depth to ensure that it had duly admitted water. One example of this apparatus

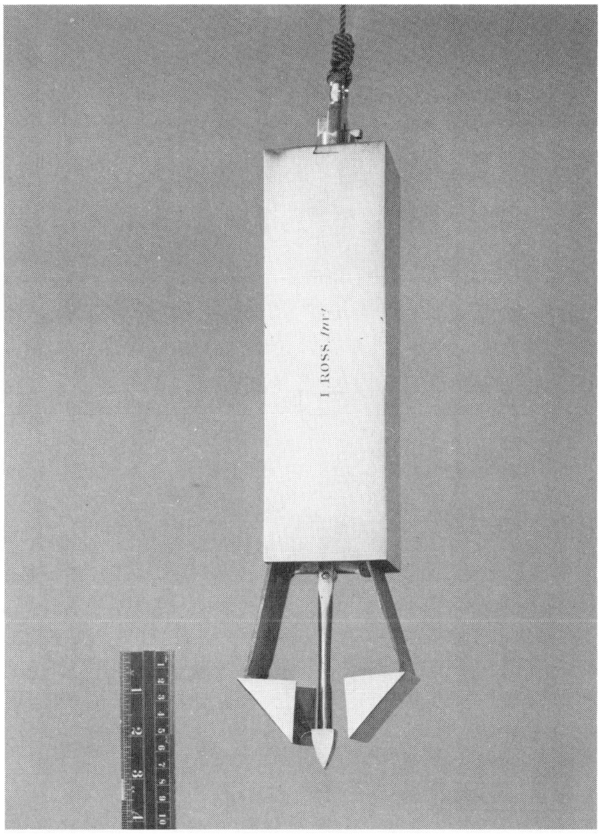

Figure 37. A half-size brass model of Ross's deep-sea clamm made by the Admiralty.

was constructed and sent out with Parry on his expedition in 1819. Ross gave its dimensions as 18 inches long and weighing 78½ lbs, holding a mere three pints of water. Parry did not record its use in his account of the voyage.

Ross was pleased with his clamm; it appeared capable of recovering samples from gravelly, muddy or rocky sediments and provided him with a means of measuring sea-bed temperatures, although he did admit that it was hard to tell when the instrument had reached the sea bed. On whale line there was a tremendous weight of rope in the water and it presented a considerable cross section to any sub-surface current. At the end of August 1818 Ross noted:

> About six it fell nearly calm for a short time, and we sounded with the deep sea clamms, which brought up a considerable quantity of mud, in which were five worms of a species that had not been seen before. There were only 650 fathoms of line out, consequently there could

Figure 38. Left, Ross's hydraphorus; *right*, Ross's deep-sea clamm. From Ross 1819b *A Description of the Deep-sea Clamms, Hydraphorus* . . . Figs 1 and 2.

not be more than that depth of water; but there might have been much less, which was probably the case, as the swell was so great, that it was uncertain, after 200 fathoms, when the machine reached the bottom.[25]

On a later occasion the sediment was so soft that the clamm was completely buried and required considerable force to extract it. Since the line was seen to run vertically down from the ship he assumed that, on starting to haul in, the 1050 fathoms paid out represented the true depth. However, when the line came on board a starfish was found attached to it at about 800 fathoms. The same had occurred a few days previously when, sounding in what he believed to be 1000 fathoms, Ross had found a brittle-star clasped round the line at 800 fathoms. Ross failed to realise that these were bottom-dwelling creatures and that, by implication, over 200 fathoms of line must have been lying on the sea bed.[26]

Describing the care with which he had made the first of these soundings Ross wrote:

The instrument took 27 minutes to descend the whole distance. When at 500 fathoms, it descended at the rate of one fathom per second, and when nearly one thousand fathoms down, it required one and a half seconds per fathom. Although the check made to the motion of the line when it struck bottom, was evident to all, I wished to put the fact beyond doubt; and for this purpose it was so nicely set as to act on the least resistance. The self-registering thermometer was then attached to it, and it was let down, first to 500 fathoms, and afterwards to 600, 700, 800 and 1000 in succession. At each time it came up empty, the thermometer each time showing a lower temperature; thus proving that the water became colder as it became deeper, and also indicating that the instrument had not reached bottom, even at the depth of 1050 fathoms. It occupied one hour for all hands to pull it up from that depth.[27]

To increase the general knowledge about the Arctic seas Ross had been instructed to throw a float overboard each day that he found himself within a surface current. The Admiralty had provided papers printed with a request in six languages that the finders of the floats should remit the paper to the Admiralty with details of the place where it had been picked up. Ross wrote his own latitude and longitude on each paper and put it into the float, which was then soldered tight and cast into the water.

Meanwhile, Lt Franklin on board HMS *Trent* had been issued with a waterbottle similar to that rejected by Phipps and Irving in 1773—a cylindrical container with conical valve above and flap-valves below (figure 39). He also took the first bottle designed by Alexander Marcet, a chemist interested in the composition of sea water in different parts of the ocean. Marcet realised that unless it could be hauled smoothly upwards, this type of sampler was always likely to re-open in rough surface water, so that its contents became mixed. To prevent this locking valves were therefore essential (figure 40, *left*):

Figure 39. Admiralty waterbottle with flap valves in the base and a conical upper valve—probably the type used on HMS *Trent* in 1818.

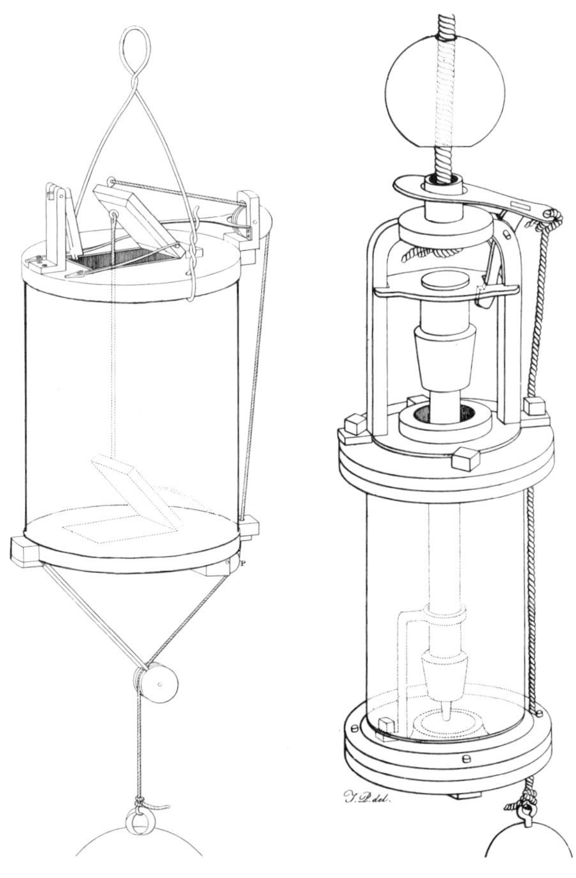

Figure 40. Marcet's waterbottles. *Left*, first bottle, 1818; *right*, second bottle, 1819. From Marcet 1819 *Phil. Trans.* **109** plates 11 and 12.

> I procured a strong cylindrical brass vessel, capable of holding about half a pint of water, and having an aperture both top and bottom, with a valve at each aperture opening upwards. But these valves, instead of being allowed to move backwards and forwards freely upon their hinges, were, when closed, firmly locked by springs; and when open, could only be kept in that way by the application of force. This force consisted in a weight of several pounds, suspended to the extremity of a cord, the other end of which was fastened to the valves, and the machine was so constructed that the moment the valves were relieved of the weight which kept them open, they closed instantly, and were forcibly locked by the above contrivance.[28]

If the bottle failed to work, then it came up still open, so there was no doubt as to whether its contents had been contaminated on the way up. Marcet asked the instrument-maker Newman to devise a means of collecting a mid-water sample and this was achieved just in time for the new model to be put on board Parry's ship, the *Hecla*, which was then about to sail for Baffin Bay on the 1819 expedition.

Newman changed the form of the bottle; the flap-valves were replaced by ground conical plugs, sliding on a central guide (see figure 40, *right*) held open by a hinged lever, counterpoised by a suspended weight. If the bottle was to sample at the bottom, then the release of the counterpoise as it struck the sea bed allowed the valves to drop shut under their own weight. To sample in mid-water, a heavy weight was sent down the line, and this overcame the pull of the counterpoise so that the valves closed. This is perhaps the earliest successful use of 'messenger release' operation of an instrument hanging below the surface.

Marcet also referred briefly to other types of bottle. That of Smithson Tennant, used in the Mediterranean, was a small wooden box with a sprung hinged lid 'like a snuff-box'. It was kept open by a wedge which was pulled out by a small turning flywheel as the box was raised. This bottle therefore predated John Ross's vane-driven hydraphorus but its wooden construction suggests that it was not used very far below the surface.

Within Davis Strait, Ross had been able to map discoveries made by Baffin in 1616 but not charted and hence long forgotten. In August 1818 he arrived in Lancaster Sound where, despite whatever he might have learned from his deep-sea surveys, Ross took a decision which was to terminate the voyage abruptly: he discerned, stretching across the waters of the Sound, a range of mountains. Some of his officers wondered if these might not be one more example of the frequent Arctic mirages and optical effects so common over ice. Parry, on board *Alexander*, was far behind, and Ross would not be contradicted. He dubbed the range 'Croker Mountains' in honour of the First Secretary of the Admiralty, and promptly set sail for home.

The voyage of Buchan and Franklin on HM Ships *Dorothea* and *Trent* was a total failure. Their ships were battered by gales off the west coast of Spitzbergen and they were forced to take shelter. On emerging, they were continually beset by pack ice, so that further progress was impossible and finally they too returned, having accomplished nothing. Scoresby had never held out much hope for the discovery of the northwest passage, and considered it unlikely that *Dorothea* and *Trent* would pass 83°N.[29] Nevertheless he watched out for the ships while he was in the same area, but did not sight them. However, that autumn he and Banks were again corresponding about the quantity of ice around Greenland. Much had broken away from the shoreline but was being drifted into new packs by winds and currents and presented a hazard to shipping.

4.4. EDWARD PARRY'S THREE ARCTIC EXPEDITIONS (1819–24)

The failure of both parts of the 1818 expedition brought strong criticism of John Ross—who was suspected of creating an excuse for an early return—and of the Navy itself. To remedy matters the Admiralty decided to despatch another expedition and in 1819 sent Edward Parry north with HM Ships *Hecla* and *Griper*. Both ships were provisioned for two years. The main objective was still to discover the northwest passage, but

> . . . the improvement of geography and navigation, as well as the general interests of science, were considered as of scarcely less importance. For this purpose, a number of valuable instruments were furnished to each ship; and Captain Sabine, of the Royal Artillery, who was recommended by the President and Council of the Royal Society, was embarked on the *Hecla* as Astronomer to the Expedition.[30]

Sabine's duties were far wider than his official title as

Astronomer implies, for he carried out or supervised many of the magnetic, gravity, meteorological and oceanographic measurements.

On his first voyage Parry sailed west through Lancaster Sound, across the charted position of the supposed 'Croker Mountains', through Barrow Strait and into Melville Sound. In September his ships passed longitude 110°W and his crews became eligible for the Parliamentary reward offered in 1818. Beset in the ice they passed the winter a little further west, but there their good fortune ended. By the following August, 1820, the pack ice was still 50 feet thick and it was clear that further westerly progress was impossible. Parry therefore made a second attempt on the passage from a more southerly approach leading out of Hudson Bay. Charts were gradually filled in but the ships were not able to break through to their previous course and thus connect the two mapped areas.

The instruments issued included one Marcet bottle for each ship and, for *Hecla* only, four Six thermometers with iron cases for their attachment to the lead-line. For sounding and sampling there were the deep-sea clamms invented by John Ross the previous year. Parry was instructed to carry out the same investigations as before: measurements of tides and currents, of the depth of the sea and the nature of its bottom sediments.

To test the surface currents, and also in the hope of thereby sending news of his progress, Parry was to cast overboard sealed bottles giving the date and his position, and this was done each day except for those days when the ships were beset in the ice pack. Measuring the velocity of currents presented a greater problem, however:

> The usual mode of trying the set and velocity of currents, by mooring a boat to a loaded iron pot let down to a considerable depth under water would appear to be a very unsatisfactory one . . . it is not easy to say how low such a current may extend and to what depth the iron pot must be sunk in order to ensure its not drifting with the boat which it is intended to moor. It may be better, indeed, to try the current in this way, when in deep water, than not to try it at all; but it is evident that a considerable current may exist, when none can thus be detected.
>
> When the water is smooth and moderately deep, perhaps under 200 fathoms, the tide or current is satisfactorily ascertained by a small boat moored to the bottom by a heavy deep sea lead, or by the deep-sea clamms . . .
>
> This mode was always adopted when practical in the course of the voyage.[31]

The deep-sea clamms also gave ambiguous results:

> It fell calm on the morning of the 2nd [August 1819], and at 9 am we sounded with the deep-sea clamms, and found 1050 fathoms by the line, on a bottom of mud and small stones, but I believe the depth of water did not exceed eight or nine hundred fathoms, the ship's drift being considerable on account of the swell. It should be remarked also that where soundings exceed five or six hundred fathoms, even in very calm weather, the actual depth must, in the usual way of obtaining it, be a matter of some uncertainty, for the weight of line causes it to run out with a velocity not perceptibly diminished, long after the lead or the clamms have struck the ground.[32]

Parry was right to suspect his results and was far more critical of them than John Ross had been. Nevertheless, he was under orders to sound whenever possible and calm days were few and far between. There was little he could do to discover the strength of any sub-surface current and to allow for its set, or to prevent the ship's drifting as the line ran down, so that it was carried away from the point at which the lead had entered the sea. It was all too common for the heavy weight to pass through the water and penetrate a soft sediment with no appreciable change in velocity and without any impact being felt on the line. Earlier in the voyage Parry had sounded where, in 1776, 330 fathoms had been reported, but he lowered 1020 fathoms of line without finding bottom. In September 1820 he let down even more line:

> In the afternoon the deep-sea clamms were sent down to the bottom with 2010 fathoms of line, which were 58 minutes in running out, during which no perceptible check could be observed, not even any alteration in the velocity with which the line ran out. In hauling it in again however, which occupied both ships' companies above an hour and a half, we found such a quantity of the line covered with mud as to prove that the whole depth of water was only 809 fathoms, the rest of the line having contrived to run out by its own weight, after the instrument had struck the ground.[33]

Three days later Parry failed to find bottom with 897 fathoms of line, and yet again in the same month with 770 fathoms run out. On such occasions the decision to run out more line or to call a halt must always have been made with an eye to the weather and state of the sea, as well as to the time and manpower required to haul in. Even 100 fathoms of line down acted as a brake; if a wind got up the vessel could not make sail or get steerage way except by cutting the line adrift, an action certain to incur the disapproval of their

Lordships at the Admiralty. There is no mention of Massey's sounder being available on these voyages, although Massey's log was streamed.

Parry referred to the Marcet waterbottle on several occasions,[34] which by all accounts worked satisfactorily down to 300 fathoms (figure 41). It was first used soon after the voyage began, a sample being collected from 100 fathoms near Rockall, for specific gravity determination. Sometimes its contents were tested for temperature, and this figure was taken to indicate that of the water at the level sampled. Though not strictly an insulating bottle, its heavy iron construction was probably bulky enough to prevent the contents altering much, provided that the apparatus was hauled up rapidly and if the temperature difference between water and air was not too great. On certain occasions Parry had to rely on such measurements for he experienced problems with his Six thermometers and had considerable doubts about their accuracy:

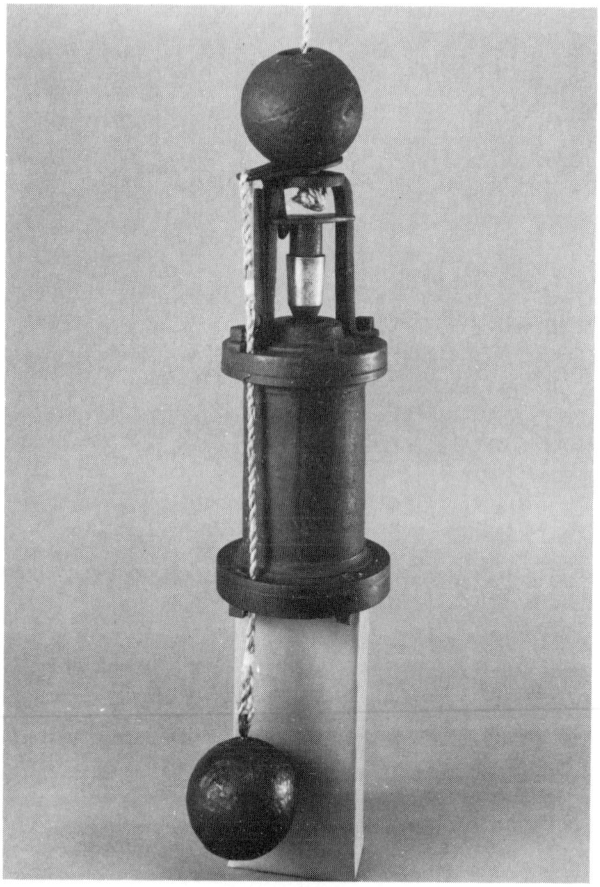

Figure 41. Marcet's second waterbottle, 1819.

> Six's thermometer was sent down to 120 fathoms, but did not indicate the temperature, owing to the mercury rising past the index, instead of pushing up before it; a failure I have often had occasion to regret in this useful instrument, when thus exposed to a very sudden change of temperature. It might perhaps be improved for this particular purpose by making the lower end of each index a little larger, so as to prevent the passage of mercury between it and the tube.[35]

This malfunction of the thermometer was caused by the sharp contraction of the glass tube when chilled, before the alcohol within had time to withdraw, compressing the contents to such an extent as to force the mercury sector past the index.

In August 1820 while sounding with the clamms, a Six thermometer was let down to 800 fathoms and left there for 2¾ hours. It registered 27°F, the surface temperature at the time being 31°F, and Parry expressed doubts about the result:

> The instrument with which this experiment was made had been used a good deal for this purpose, and did not, perhaps, indicate the temperature with very great accuracy.[36]

The indexes, held against the side of the tube by a bristle, became loose in time as the bristle lost its springiness in the alcohol.

Despite the failure to break through to a polar ocean, the first voyage was judged a success. The crews were in good health, both ships survived the battering by ice, and the geographical discoveries encouraged the whaling fleets to extend their activities into the rich waters of Lancaster Sound. Even if the seaway that Parry had entered led only to an impenetrable barrier, there was still hope that a parallel route might be found further to the south.

In the following summer of 1821, HMS *Hecla* and a sister ship *Fury*, went north under Parry's command and again overwintered in the Arctic. Again a seaway was discovered but was blocked by ice, and after a second winter during which his crews suffered badly from scurvy, Parry returned home. The expedition made several overland treks which brought them into contact with Eskimos. Relatively little sailing was done, although some oceanographic work did prove possible. Many more instruments were carried on this voyage and included Six thermometers (six for *Fury*, two for *Hecla*) and two Marcet waterbottles for each ship. Parry described how the sea water temperatures were measured:

> The temperature of the sea below the surface was sometimes obtained with Six's self-registered thermometer attached to the deep-sea lead; but more commonly (in consequence of the frequent failure of that instrument when exposed to sudden changes) by bringing up water in the bottle contrived by Dr Marcet . . . This simple and useful apparatus was now somewhat improved by a strong spring enclosed within the box and obliging the bolt in whatever position it might be placed, to close the apertures as soon as the catch was released, instead of trusting to its own weight as before.[37]

As on the previous voyage, the bottle was used to depths around 300 fathoms, with apparent success.

Parry sailed again in 1824 on board *Fury* and *Hecla*, but the ice conditions were bad and no westerly progress was made. The crews overwintered, carrying out regular magnetic and meteorological observations, but the following August the *Fury* was damaged and while repairs were being attempted she was finally wrecked, so that both crews had to return on *Hecla.* Parry had been appointed Hydrographer of the Navy in 1823 and his place in Arctic exploration was taken by Franklin, whose endeavours had been mainly overland or in small boats.

One last expedition falls within this time period, however. With private backing, John Ross sailed into the Arctic in 1829 on the paddle-steamer *Victory,* which gave the crew a great deal of mechanical trouble. To send such a novel form of transport into a region where there was no coal for refuelling, no facilities for repair other than those on the ship herself, and where climatic deterioration was manifest, would be incomprehensible today. On the *Victory* Ross was beset for three winters and finally he had to abandon ship while his crew made their way by boat to Lancaster Sound where they were picked up by Ross's old ship, the *Isabella,* now a Hull whaler. In his narrative Ross devoted most space to the difficulties of operating a steamship and to his encounters with Eskimos, but he did refer to the fate of some of his instruments:

Figure 42. 'Deep soundings; or, no bottom with 4600 fathoms'—sounding from boats during the Antarctic expedition of 1839–43. From Ross 1847 *Voyage . . . in the Southern and Antarctic Regions* vol. 2 p355.

The deep-sea clamms, Dr Marcet's waterbottles, Massey's patent log and other instruments of minor importance . . . were left at Victoria Harbour.[38]

John Ross's nephew, James Clark Ross, sailed with his uncle and on other Arctic voyages, learning the various techniques that the Admiralty was coming to demand of any commanders appointed to lead scientific voyages. He commanded HM Ships *Erebus* and *Terror* in 1839–42, taking them down to explore the fringes of the Antarctic continent as part of the Colonial Magnetic Observatories' programme.[39] Figure 42 shows his boats' crews trying—and failing—to get a deep sounding in that great southern ocean. It leads us from the era of the famous polar voyages and their pioneering oceanographic work to the next chapter which deals with the trials and tribulations of getting accurate deep soundings in any ocean.

REFERENCES AND NOTES

1. Phipps C 1774 *Voyage Towards the North Pole, 1773* (London)
2. Phipps C 1774 *Op. cit.* p27
3. Phipps C 1774 *Op. cit.* p32
4. Bouguer P 1747 'Sur une nouvelle construction de loch' *Hist. Acad. R. Sci. Paris* **60** pp96–102; *Mém. Acad. R. Sci. Paris* **60** pp644–64
5. Phipps C 1774 *Op. cit.* Appendix, p144
6. Scoresby W 1820 *An Account of the Arctic Regions* (2 vols) and 1822 *Journal of a Voyage to the Northern Whale Fishery* (Edinburgh)
7. Stamp T and Stamp C 1976 *William Scoresby, Arctic Scientist* (Whitby)
8. Scott H W (ed) 1966 *Walker J, Lectures on Geology &c.* (Chicago)
9. Walker J 1792 *Institutes of natural history, containing the heads of lectures in natural history delivered . . . in the University of Edinburgh* (Edinburgh)
10. Scoresby W *c*1809–10 *Natural History Notebook,* Mss Scoresby Archive, Whitby Museum
11. The spelling of Cary's name is variable.
12. Scoresby W 1811 Letter to Jameson, 21 October 1811, Mss Scoresby Archive, Whitby Museum
13. Scoresby W 1817 Letter to Jameson, 27 August 1817, Mss Scoresby Archive, Whitby Museum
14. Scoresby W 1854 'On deep-sea sounding and the errors therein from strata currents, with suggestions for their investigation' *Rep. Br. Ass. Adv. Sci.* (Sections) pp22–26
15. Scoresby W 1820 *Op. cit.* pp196–7
16. Perkins J 1820 'On the compressibility of water' *Phil. Trans.* **110** pp324–9
17. Ross J 1819a *A Voyage of Discovery . . . in HM Ships 'Isabella' and 'Alexander'* (London) Official Instructions, pp1–4
18. Davy H 1818 'A description of the apparatus . . . for bringing up water from certain depths in the sea' *J. Sci. Arts* **5** pp231–3
19. Ross J 1819a *Op. cit.* Appendix X, p cxxix
20. Prestwich J 1874 'Tables of temperatures of the sea at different depths beneath the surface . . .' *Phil. Trans.* **165** pp587–672
21. Rice A 1975 'The oceanography of John Ross's Arctic expedition of 1818: a reappraisal' *J. Soc. Bibl. Nat. Hist.* **7** pp291–319
22. Ross J 1819a *Op. cit.* pp cxxxiii–cxxxvi
23. Ross J 1819b *A description of the deep-sea clamms, Hydraphorus and Marine Artificial Horizon* (London)
24. Ross J 1819b *Op. cit.*
25. Ross J 1819a *Op. cit.* pp175–6
26. Rice A 1975 *Op. cit.*
27. Ross J 1819a *Op. cit.* pp193
28. Marcet A 1819 'On the specific gravity, and temperature of sea water, in different parts of the ocean and in particular seas, with some account of their saline contents' *Phil. Trans.* **109** pp161–208
   Marcet mentioned that he had become interested in the construction of an efficient bottle after seeing in an instrument-maker's shop window the bottle used by Irving. From his description, however, it is not the insulated bottle that he is referring to, but rather one like that shown in figure 39.
29. Scoresby W 1818 Letter to Jameson, 28 January 1818, Mss Scoresby Archive, Whitby Museum
30. Parry W E 1821 *Journal of a Voyage for the Discovery of a Northwest Passage . . . Performed in 1819–20 in HM Ships 'Hecla' and 'Griper'* . . . (London) Introduction, pvi
31. Parry W E 1821 *Op. cit.* Appendix VI, p cxlvi
32. Parry W E 1821 *Op. cit.* p30
33. Parry W E 1821 *Op. cit.* p293
34. Parry W E 1821 *Op. cit.* pp4, 45 and 295, for example
35. Parry W E 1821 *Op. cit.* p4
36. Parry W E 1821 *Op. cit.* p293
37. Parry W E 1824 *Journal of a Second Voyage for the Discovery of a Northwest Passage Performed in the Years 1821–23 in HM Ships 'Fury' and 'Hecla'* (London) Introduction, p xvi
38. Ross J 1835 *Second Arctic Voyage in Search of the Northwest Passage* (2 vols) (London) Appendix, p cviii
39. Ross J C 1847 *A Voyage of Discovery and Research in Southern and Antarctic Waters, 1839–43* (London)

Chapter 5

The Grand Victorian Technology: Submarine Telegraphy

5.1. TELEGRAPHY: THE POLITICAL AND ECONOMIC IMPLICATIONS

During the first half of the nineteenth century the Royal Navy's scientific endeavours had been directed in particular to the polar regions, but physical oceanography took second place to geophysics. Elsewhere they undertook hydrographic surveys, but paid scant attention to the deeper parts of the ocean; the few sites sampled were too widely separated and the apparatus used too unreliable for notable advances to be made. The Navy suffered from Parliament's reluctance to allocate adequate funds during peacetime and, in addition to its commitments in war and hydrography, also had to supply vessels for fishery protection and for excise and coastguard duties.

By the middle of the nineteenth century the greatest Victorian technology (as some considered it) became feasible: a telegraph system girding the world. For the first time in history, ambassadors and colonial governors suddenly lost their autonomy and were reduced to the status of spokesmen, referring all major decisions to their seat of government. Masters of trading ships had to consult their owners before buying and selling cargoes since world market prices were immediately telegraphed to their head offices. Notice of the outbreak of war and its consequent progress could be made available for ships calling at foreign ports while world news was published in local newspapers within hours of the event. On a smaller scale, weather reports transmitted from distant stations to a central point contributed to a storm-warning service, at first restricted to Britain and France, but soon to expand, and this undoubtedly resulted in the saving of many lives at sea.

To bring about such changes the telegraph lines had to cross not only mountains, moorland and desert, but also the bed of the sea. After many miles had been laid both in North America and in Europe the cable went to sea, so to speak, in the shallow waters of the English Channel. In 1840 Professor Wheatstone had first suggested to Parliament that a cross-Channel link should join British and French telegraph systems. A first attempt made in 1850 proved a failure but a cable the following year performed better, and soon a transatlantic cable was being planned. Laying commenced in 1857 and was completed the following year. By 1860 a network of cables across the continental shelf linked Britain and Europe to their offshore islands.

Before that came to pass however, the Crimean War, which started in 1854, benefited from messages passed via cables laid through the Black Sea for that period. In 1858 the British government took over the last of the Indian colonies formerly controlled by the East India Company with the confident prospect of establishing telegraphic communication over several land and maritime routes between Britain and the regions of that vast sub-continent.

Crossing the Atlantic was a daunting proposition, however. Three thousand miles of cable had to be laid in water up to 15 000 feet deep. Mid-ocean staging posts could be established on the Azores and Madeira for some crossings, but not for a direct London–New York line. Coastal routes passed over ground made hazardous by strong tidal currents or sharp rocks, or even subjected to volcanic activity as in parts of the Mediterranean. The active life of such cables was often brief so that cable ships had also to pull up and repair fractured lines.

The Pacific Ocean was too vast to be tackled in a single stage except in the north, so it was eventually spanned as a series of inter-island hops. The new wireless transmission which came into use early in the twentieth century led to a gradual reduction in the annual increase of cable mileage in service. However, one disadvantage of wireless was its open broadcast, for all to listen in. The disadvantage of the cable network was its vulnerability to attack where it ran through lands subject to tribal unrest. It was to avoid these troubles that the Indian cable was laid in the Red Sea, rather than overland along its shore.

The British market was buoyant in North and South America; the importance of trading links with the Far East and India, and the commerce between European nations was encouraging. Occasional failures damped neither the optimism of telegraph engineers, to whom no problem was insuperable, nor the willingness of investors to put their money into any cable enterprise which seemed likely to yield dividends. Under such pressures, many lines were laid carelessly or in unsuitable regions and there were failures enough, though the mileage continued to increase steadily.

Under the stimulus of the demand from the developing telegraph industry, three important technical advances were made. The first was the ability to manufacture lengths of cable to a guaranteed standard with respect to the electrical capacity of the copper core and its insulant, and to the mechanical strength of the outer iron armour. The second advance was the development of the steam engine, which was essential for holding the laying vessel on a predetermined course and speed, irrespective of the wind and weather, and also for operating the paying-out winches so that at any given moment the weight of cable hanging in the water did not exceed its own breaking strain. The third advance, to which this chaper will be devoted, was the ability to provide an accurate sea-bed survey along the proposed route.

Closely spaced soundings and bottom samplings had to be made along a track fixed precisely by astronomical observations. Topography dictated the mileage of cable to be manufactured for, when laid, it had to rest on the bottom and not hang unsupported from ridge to ridge. The total length required therefore exceeded the direct coast-to-coast distance and thus increased the amount of capital needed to be raised from investors. The survey had to detect areas underlain by rocks or swept by currents so that alternatives could be sought, otherwise heavy armouring was put on the cable where it was exposed to such hazards and it was common practice to lay a heavier cable across inshore ground, reducing its diameter as it entered deeper and calmer water.

Temperature measurements and water samples were also taken, for the sea bed temperature affected the electrical resistance of both core and insulant, while the water might contain some corrosive element likely to damage the armour. For the first time oceanographic apparatus was brought to the attention of electricians, engineers and other inventors. They went to sea with the Navy, who were mostly responsible for surveying routes until the 1870s on behalf of British companies. The problems involved in the design and use of such apparatus, and the solutions that they offered, both practical and crazy, form the subject of this and following chapters.

5.2. PRE-1850 DEEP SOUNDINGS AND THE BROOKE SOUNDER (1853)

By 1850 the Americans were leading the British in the development of deep-sea sounding techniques, but there were problems. The dilemma was this: the use of a heavy sinker to carry a line down through several miles of water often caused the line to break when it was hauled up; using a light lead and line made it difficult to tell when bottom was reached, while friction between the line and water tended to prevent the lead reaching bottom at all. Unless a sample was recovered there was no proof that bottom had been reached, and it was likely that the line paid out greatly exceeded the vertical distance between surface and sea bed. Waywiser sounders, as we have seen, were impractical in deep water.

In 1850, however, a solution was offered by Lieutenant Matthew Maury of the National Observatory in Washington. He recommended the following technique: a 32-lb shot attached to light twine was lowered from a boat, but this line was doubled over the first part of its length to prevent breakage as it passed through the rough surface water. The rate of descent was timed, and when it decreased markedly the shot was judged to have reached bottom and the twine was cut, the distance paid out being taken as the true depth. The method was rapid, fairly cheap, easy to operate, and Maury was at first confident that it was reliable. He too had designed a waywiser sounder to check the results but found it worked no better than others of this type.[1] However, as the number reports grew, it was apparent that the method was not as good as had been hoped because the enormous lengths of line that were paid out, sometimes near to proven shallow water, could not represent a true depth.

Trials were made with wire line as early as 1834:

> In April last, a trial of a new sounding machine was made off Scilly in HM steam vessel *Medea* . . . the invention of Mr Read, a gentleman from the United States. The lead, formed of two parts, was fastened to the end of a copper wire . . . and when at the bottom, one of the parts was to be disengaged. An index on the reel showed the number of revolutions it had made, while the lead descended and measured the depth.[2]

The experiment proved a failure. In 1848 Captain Barnett, surveying in the North Atlantic in HMS *Thunder*, tried a 61-lb weight on an iron wire, but he too was unsuccessful, since the wire broke at 2000 fathoms.

In 1851–52 Lt Samuel Lee took the US brig *Dolphin* into the Atlantic under orders to sound at regular intervals on each crossing and, in particular, to sound at the reported positions of some 56 'vigias'—those rocks, reefs or shoals whose existence was in doubt. Lee took on board Stellwagen cup leads—deep-sea leads with a small cone at the base to collect some bottom sediment samples—and Burt's buoys, and for the deeper water where he could not hope to recover his apparatus he received 225 000 fathoms of flax twine and 10 000 fathoms of triple line. He did manage to sound to 3000 fathoms, but throughout the voyage he had to contend with two handicaps: the weather, which made it impossible to hold the line vertical over its initial point of descent, or caused it to break during the early stages of the operation; and the poor quality of the twine itself. Lee found that it contained many rotten or weak portions and frequently it stretched or broke when a considerable length was lowered into the water.

The weather also made it difficult for Lee to carry out the other part of his instructions, which was to rate the currents at and below the surface. He was in addition to measure water temperature, salinity and transparency; since he had no serviceable deep-sea thermometers, Lee sent down a standard thermometer inside screw-top glass or iron bottles, but found that these leaked in deep water. He was also provided with a wooden and a composition-metal, valved water sampler. Both were intended to insulate the water sample but did not in fact do so, and in any case were themselves lost during the course of the voyage when their sounding line gave way. Lee's report made plain the problems that beset him, and showed how open to question some of his results were.[3]

When using thin line, whether cord or wire, it was difficult to lower it smoothly through the first few fathoms and to avoid breakage as the ship pitched. Wire and fibre had different densities when immersed in water, and so ran out at different rates as the line entered the sea. In addition, wire ran out faster and continued to do so after the sinker had come to rest at the sea bed, but fibre line was braked by friction and hence ran out more slowly, so that the decrease brought about by the sinker's arrival at the bottom might pass unnoticed. The friction could be eased by replacing the usual hemp or sisal by waxed twine or silk line.

To recover a bottom sample the sounder had to be lightened before it was raised, and this meant that the sinker had to be left at the bottom. The concept was not new, and was from time to time revived as we have seen. Trials were also made on French ships; after the sounding failures reported by Captain Du Petit-Thouars on *La Vénus* (1836–39), the French astronomer H A E Faye proposed a lineless pressure sounder with detachable spherical weights. No drawings were made public and, like most other similar inventions of its kind, it seems not to have been a practical proposition.[4]

The breakthrough eventually came from the Washington Naval Observatory in 1852, when a young naval officer, John Brooke, showed Maury drawings of a sampler weighted by a cannon ball which was cast off at the sea bed, allowing the light sampler to be hauled up alone.[5] As first designed, the Brooke sounder consisted of a rod, its base armed with tallow, fitting into a hollow cast shot.[6] The sounding line was attached to a hinged arm at the top of the rod, which also supported the cord harness in which the shot rested during its descent. At the bottom, strain came off the line, the arm dropped, slipping the harness and shot. When hauled up, these were left behind and the rod with its confirmatory sediment adhering to the tallow was safely raised.

When the first model of the sounder was made, the instrument-maker suggested that two hinged arms would be an improvement (figure 43), but in practice, unless the pivots were most carefully positioned they did not trip simultaneously and so the apparatus failed

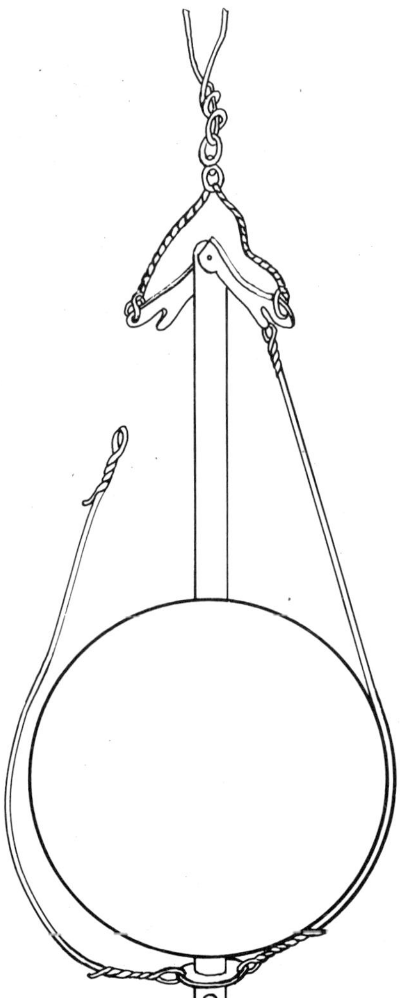

Figure 43. Brooke's two-arm sounder, shown in the act of detaching the weight prior to hauling up. From Maury 1855 *The Physical Geography of the Sea* plate III.

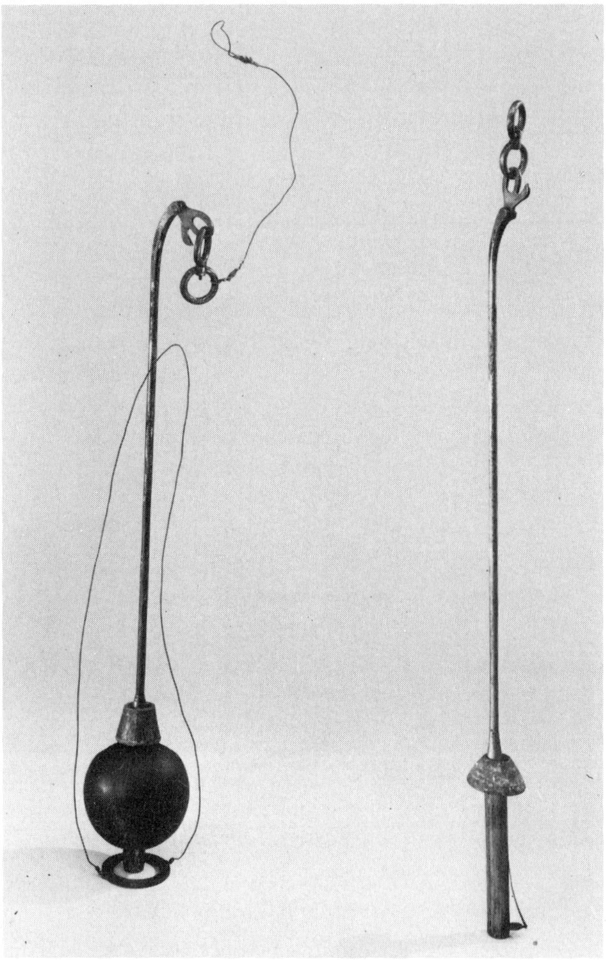

Figure 44. Admiralty modifications of Brooke sounders showing the valved sampler tube.

to detach. Another serious defect was the sounder's tendency to rotate and this prevented it being used in moderate depths with a waywiser sounder added. Brooke therefore reverted within a few years to the single-arm pattern, which had been his original design, but by then the sounder was familiar to hydrographers on both sides of the Atlantic and had inspired many imitations (figure 44).

The single-arm form to which Brooke added goose-quills to increase the collecting capacity of the sampler, was tested by Lt Berryman on USS *Dolphin* in 1853, and by the end of that year Maury issued instructions for its general use. A 64-lb shot was to be used on twine 1/10 inch in diameter, furnished on 5000 fathom reels.[7] Examples of the sounder were also sent to the British Navy for trials.

When, in August 1856, Berryman took USS *Arctic* across the Atlantic to sound for the first transatlantic cable, he used a Brooke sounder which was much altered from the original. Its tapered sinker encountered less resistance as it passed through the water, and a valve over the sampler tube prevented fine sediments being washed out on the way up.

Maury rejected the profile that Berryman obtained on this voyage because the results of the east–west run did not tally with those made in the reverse direction.[8] When the ship's journal was examined he found that insufficient care had been taken in making each sounding and locating it accurately on the chart. The operation therefore had to be repeated in the following year by the British Navy, who sent out Lt Joseph Dayman in HMS *Cyclops* on a course slightly north of the one taken by Berryman (figure 45).[9] Edward Massey constructed the Brooke sounders, incorporating modifications such as elongated sinker weights, iron stays to replace the cord harness, and valved sinker tubes, as requested by Dayman (figure 46). Multiple weights could be loaded and 20 Massey waywisers were taken, to be fixed to the sampler shaft when the water was not too deep. A variety of hemp lines, and six sets of Burt's buoy and nipper were added. Soundings could also be made from boats, using a recoverable lead on silk line. The Admiralty instructed Dayman to sound closely in water of less than 100 fathoms; every ten miles till he reached much deeper water, and then every 40 or 50 miles up to the approaches to the Newfoundland shore. Dayman also took self-registering thermometers, a Lecoëntre waywiser sounder (figure 47) used by the French for trials,[10] and he was ordered to try the surface and submarine currents if this was possible. Although satisfied that his soundings were accurate, Dayman found that the readings of the waywisers varied according to the thickness of line used, so that calibrations made with any one rope were not applicable to any other.

Figure 45. Four different profiles across the North Atlantic sea bed, showing the disparity resulting from the various methods of sounding and of position finding. From Maury 1858 *Explanations and Sailing Directions . . .* plate XII.

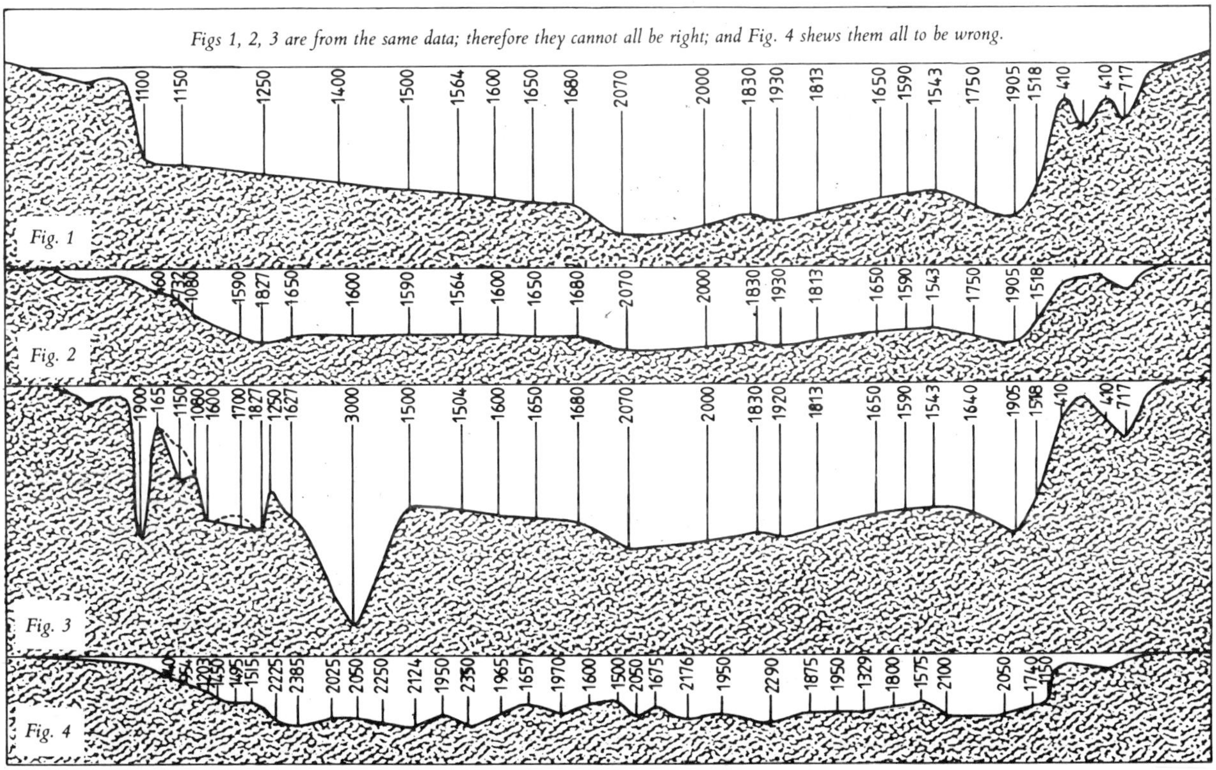

PROFILE OF TELEGRAPHIC PLATEAU *Fig. 1 Arctic's Soundings by Chart No. 1 Scale ¼ of Original*
Fig. 2 do do do No. 2 do do
Fig. 3 do do by her Abstract Log
Fig. 4 Cyclops' do from Admiralty Chart

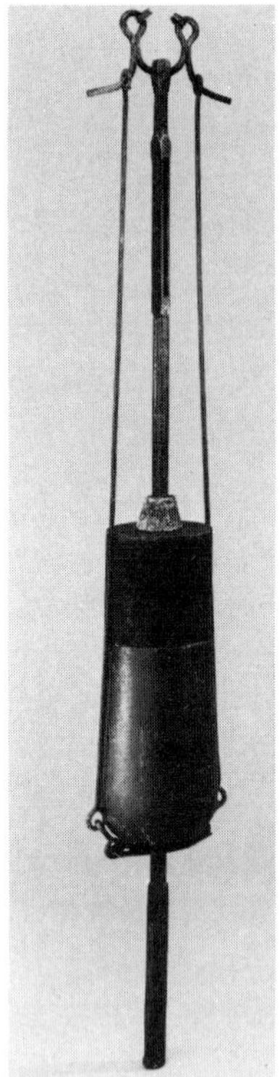

Figure 46. Dayman's sounder. Metal rods replaced the wire harness and the central shaft is slotted to accommodate a Massey sounder.

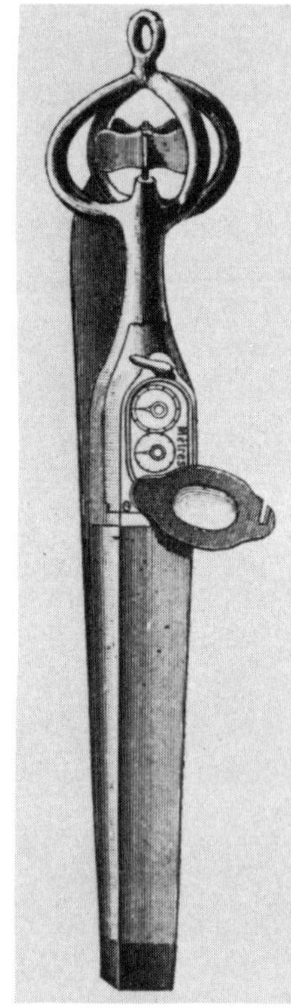

Figure 47. Lecoëntre waywiser sounder. From Richard 1907 *L'Océanographie*, Fig. 23.

5.3. VARIATIONS ON A THEME: DETACHING-WEIGHT SOUNDERS

While transatlantic surveys were progressing, Captain Spratt of HMS *Spitfire* was off Sevastopol in 1855 and in the Mediterranean in 1856–57.[11] Spratt found that a 10 or 12-lb weight attached to silk line gave him the best results in water less than 2000 fathoms deep (silk has a specific gravity close to that of sea water and remains less heavy than the sinker, irrespective of the length submerged). His Brooke sounders were made by the ship's blacksmith and suffered from the defect common to the two-armed version, namely that unless carefully made the arms did not drop together. Brooke in fact wrote to Spratt, and also to Mansell, about the need for careful positioning of these arms.[12] For sounding in the Black Sea where the water seldom reached 300 fathoms, the blacksmith Carmelo Bonnici constructed a small detaching device known as Bonnici's claws to release any weight that could be carried in a simple rope sling. Spratt used it with a small armed lead (figure 48), but in deeper water it carried a 68-lb shot for sounding alone, where a sample was not required (figure 49).

In the Mediterranean Spratt practiced several methods of sounding at each station and ultimately recommended that three operations should be made at each point where sounding and bottom samples were needed.[13] First, the submarine current should be tested; next, a sounding should be made with a heavy line and sampler, allowing ample slack to ensure that the sampler reached the bottom, and finally, an accurate measurement of depth should be made with a twine line, such as that used for sail-making. When the sounder was judged to have reached bottom, the line was tested by measuring the strain with a spring steelyard balance. This exercise often broke the line, and hence the second operation was carried out using a heavier rope.

In 1857, Commander A L Mansell also sounded in the

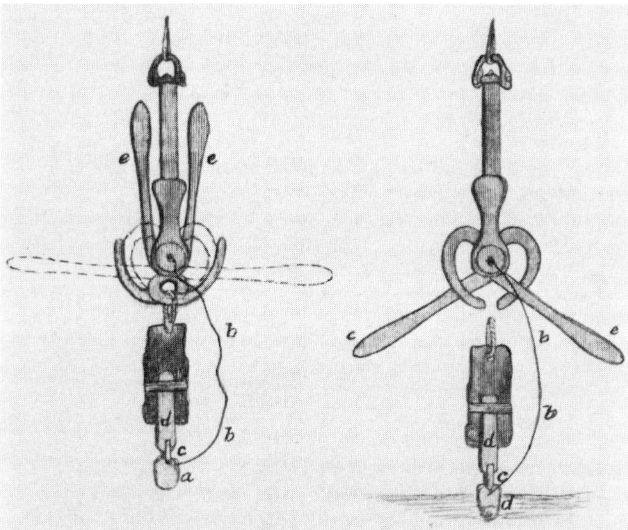

Figure 48. Bonnici's claws—the original version as used by Spratt. a is a small cup armed with tallow, fastened by a wire or line bb to the claws. Two wooden pieces d, lashed to the lead weight, have small spring hooks c to hold the cup during descent. Strain comes off the line at the bottom, allowing the arms ee to open the claws and release the weight. From Spratt 1856 *Naut. Mag.* **25** p158, Figs 1 and 2.

Mediterranean (figure 50), using a Brooke sounder weighted with a deep-sea lead (figure 51).[14] He found that the sounder penetrated so far into soft mud that the weight could not detach. He also tried Bonnici's claws but could not recover an adequate sample. His illustrations of these instruments show additional weights at the ends of the arms; it was then a small step to the development of the detacher made by his Master, Francis Skead, which was a one-armed version of Bonnici's claws with a small cup to hold sediment (figure 52).

Figure 49. Bonnici's claws, Admiralty version, with dummy weight representing a 68-lb shot. This was for sounding only: no samples were recovered.

Figure 50. Chart and sections made by soundings for a telegraph cable between Malta and Crete, made by Spratt and Mansell in 1857.

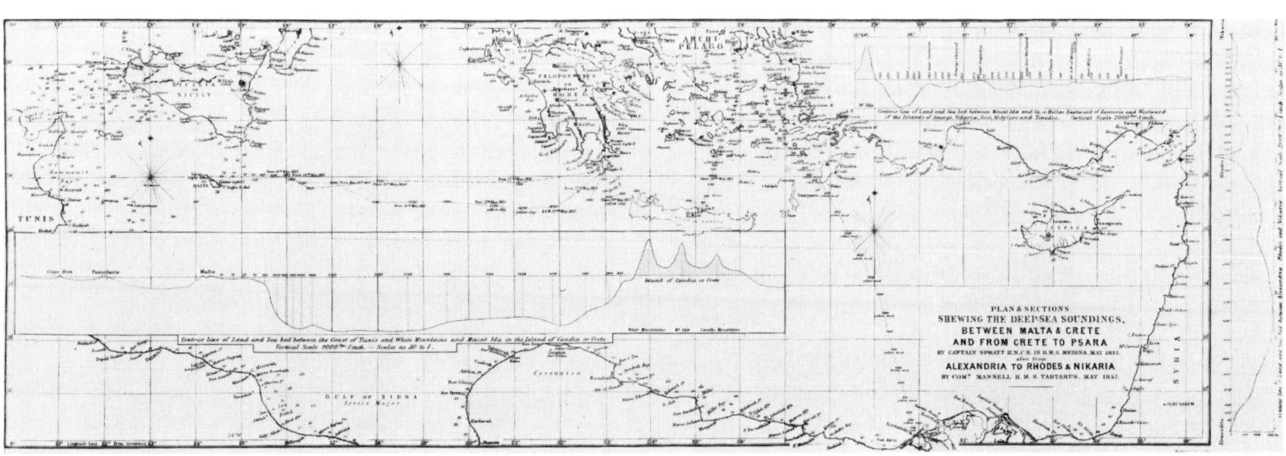

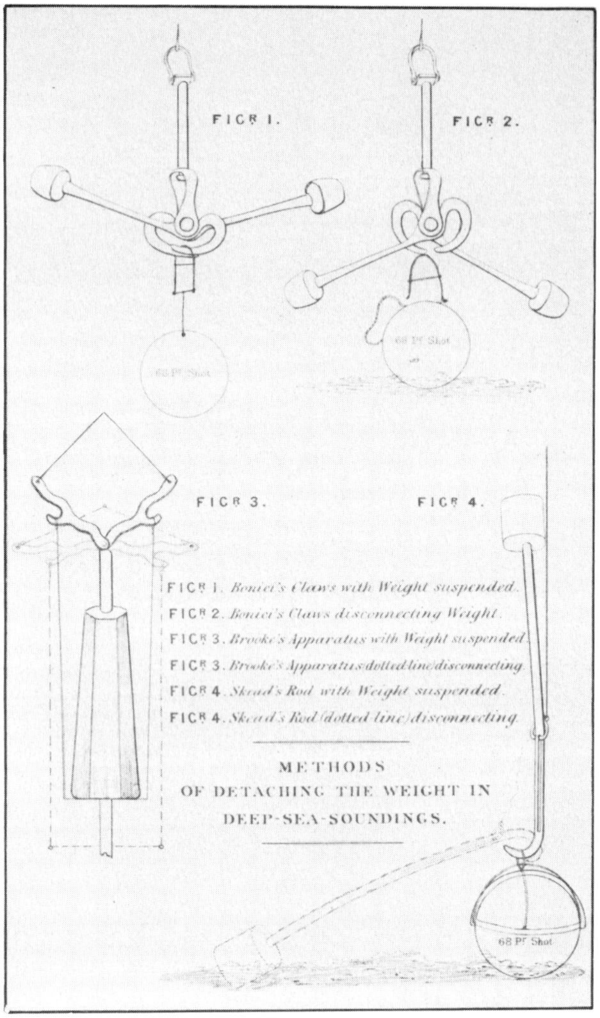

Figure 51. Four types of sounder tried by Mansell in the Mediterranean in 1857. From Mansell 1857 *Naut. Mag.* **26** p512, Figs 1–4.

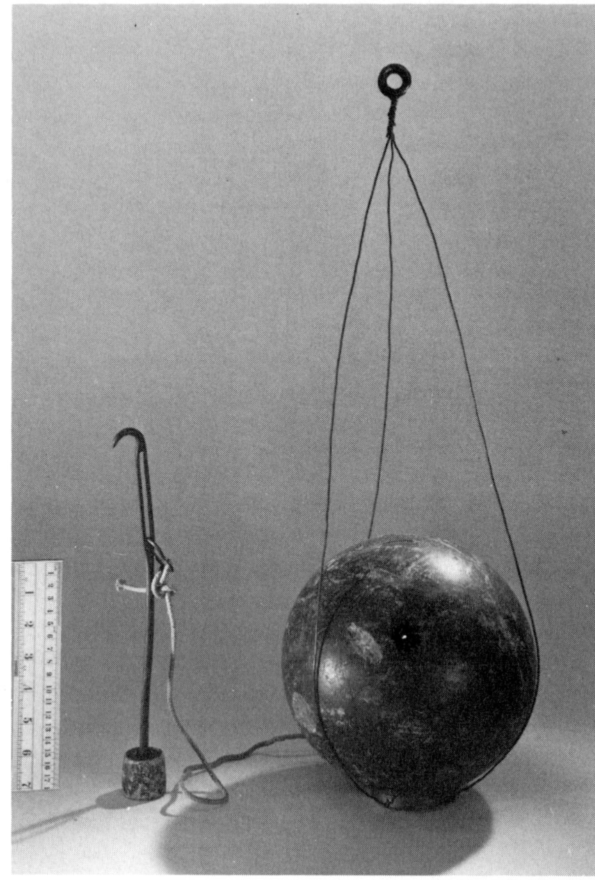

Figure 52. Skead's one-armed version of Bonnici's detaching claws: the Skead detacher and specimen cup. This is an Admiralty model with dummy 68-lb weight.

As the number of sounding voyages increased, and until about 1870 these were still undertaken by the Royal Navy on behalf of British cable companies, it became apparent that no single sounder was capable of operating at all depths and over all types of bottom. For example, if a sampler penetrated soft oozes it could fail to detach its weights, or indeed could not be pulled up at all. On gravelly sediments pebbles could jam the sampler valve, whilst on rocky ground the sampler recovered nothing and often became lodged in the rocks (figure 53).

The Atlantic cable laid in 1858 never gave good service, so that in 1860 the Privy Council appointed a committee to investigate the manufacture and laying of existing cables and to consider alternative routes.[15] Underwater stages could be shortened by taking the cable via Iceland or the Azores, but both of these routes had disadvantages. The committee enquired closely into the thoroughness of sounding surveys, and one of those questioned was Joseph Dayman whose answers[16] revealed more about his methods than appeared in his earlier published report.[17] Dayman's opinion was that the depths given by Brooke's sounder were probably too great by a factor of 150 fathoms in 2000, arising from the difficulty of estimating, from time intervals at which equal lengths of line ran out, the exact moment at which the sounder reached bottom, and also the impossibility of knowing how slack the line was. Dayman had sometimes tried another method: a very heavy weight was fixed to the line, greater than could

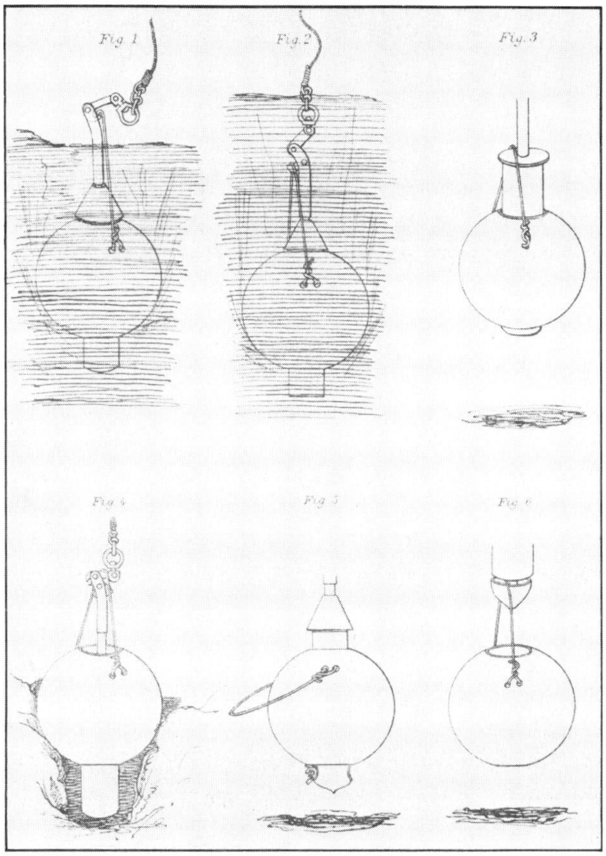

Figure 53. Causes of sinkers failing to detach. From Sigsbee 1880 *Deep-sea Sounding and Dredging* plate 4.

be lifted off the bottom at depth. When this weight rested on the sea bed, the line was hauled taut by means of a donkey engine or by hand, until it broke. This was akin to Spratt's test with the spring balance, but inevitably more costly in terms of line lost. Questioned about the advantages of wire sounding, Dayman replied that he had no experience of its use, but he understood that it was difficult to judge from its rate of running-out exactly when bottom was reached.

Maury had criticised Berryman's survey in *Arctic* on the grounds that the location of each sounding had not been noted, but merely the position of the ship each day at noon. Dayman reported that each of his soundings had been fixed by astronomical observation and by distances run by patent log, and also, when in sight of land, by bearings.

A precise depth sounding was of little value unless its position was accurately known. The longitude of some ports was still imperfectly known, until the transmission of time signals by telegraph, so that there was often uncertainty whether the laying vessel was following exactly along the route prospected earlier.

One transatlantic crossing under consideration was that via the Faeroes, Iceland and Greenland. The Admiralty undertook to survey this route and in 1860 Sir Leopold McClintock took HMS *Bulldog* along the proposed track, returning home via Greenland and Rockall.[18] *Bulldog* was a large paddle-steamer furnished with a deck engine to power the sounding reels. She carried the usual complement of leads, modified Brooke sounders and heavy weights to be used on light line to obtain true vertical depth. Each sounding took about three hours. With the advent of steam power it was no longer essential to sound from a boat as the ship herself could be kept in position near the Burt's buoy, but the business was still time-consuming and the vessel was frequently delayed by bad weather.

McClintock was disappointed by the performance of the Brooke sounder, since its valve often failed to close and the amount of sediment collected was in any case very small. For moderate depths he preferred a Stellwagen cup (figure 54)[19] but for deep-sea work he came to rely on a grab devised by his two engineers, Roughton and Steil, with the assistance of the ship's naturalist, Dr G C Wallich. The aptly named *Bulldog* sounder (or grab) scooped up superficial mud plus underlying sediment even when this consisted of stones an inch or more in diameter.[20] Working round the margin of the Atlantic basin, McClintock was finding coarser material than that usually found on the abyssal plains.

Wallich, whose duties included the care and identification of the sediments and marine fauna collected, subsequently published a narrative of this voyage in which he claimed credit for this sounder.[21] Indeed, he took out a provisional patent for it in 1860, though he failed to be granted full protection. Two versions of the grab survive; one is probably that referred to by McClintock:

> Since our return to England, Mr Steil's sounding apparatus for bringing up a fair quantity of the bottom has been made in Portsmouth Dockyard and deposited there as an exact model . . .[22]

The other was made by Wallich himself, perhaps as part of his patent claim (figure 55).

McClintock's version comprised an elongated sinker weight, released at the bottom by a modified Bonnici

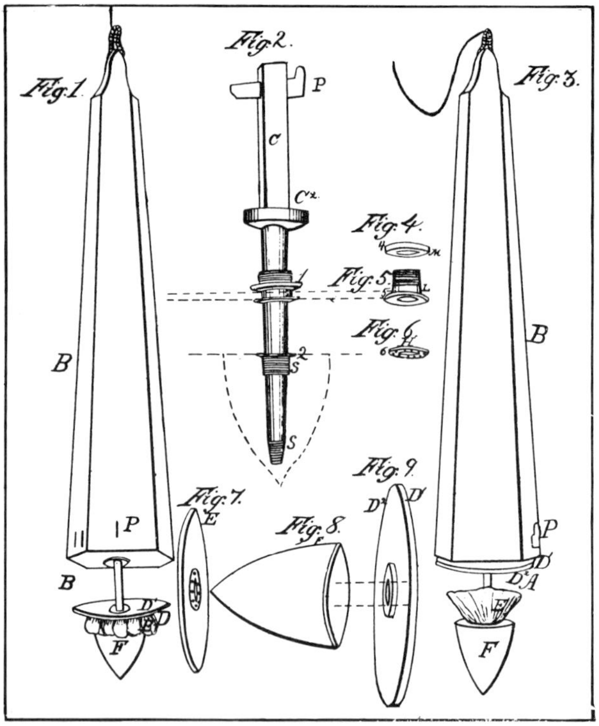

Figure 54. Patent drawing of the Stellwagen bathometer or cup lead, 1842. B, lead; P, pin; D, plate: E, soft leather; and F, specimen cup (*US Patent Specification* No. 2899).

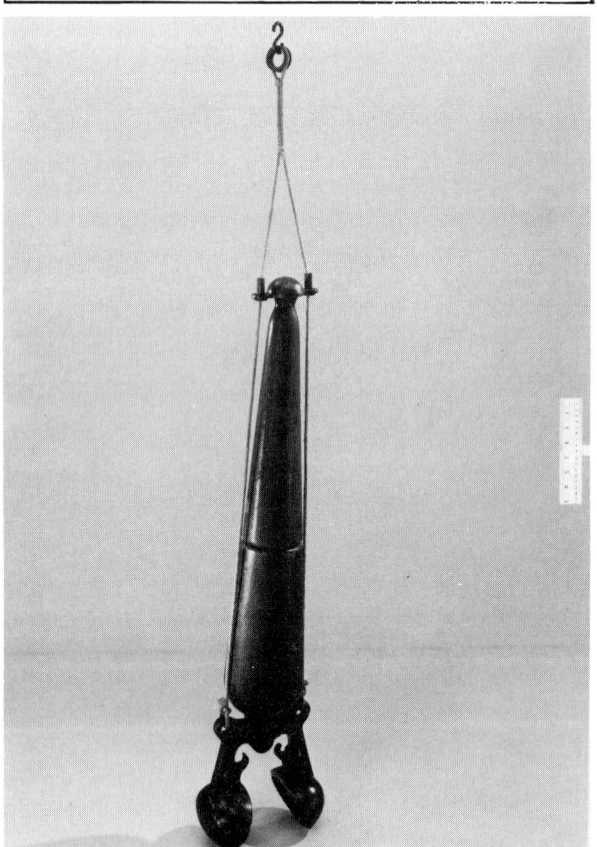

Figure 55. Wallich's model for the *Bulldog* sounder, made after the 1860 voyage.

detacher. Beneath this was the grab, whose jaws were held open until they had penetrated the sea bed. As the weight fell away, the jaws were snapped shut by a rubber band around their upper portion and they were then hauled up, closed, to the surface. Wallich's sounder differed slightly in the shape of the jaws and the location of the rubber band, but was otherwise similar.

A *Bulldog* sounder was issued to Captain Hoskyn of HMS *Porcupine* in 1862 for his detailed survey of the continental slope west of Ireland. His deepest successful sounding was 1750 fathoms, but he lost the line on most of those attempted in water over 1500 fathoms deep.[23]

Whatever its shortcomings on certain types of ground, the Brooke sounder still had its supporters. Captain Shortland, surveying in HMS *Hydra* in 1868 along the proposed Red Sea cable route, had been issued with a *Bulldog* grab but had obtained better results when using a modified Brooke sounder devised on board.[24] *Hydra*'s blacksmith Gibbs and two others each received the sum of £1 for their ingenuity in constructing a spring to throw off the sounder weight at the sea bed. The first sounder they made incorporated an old sampling tube that had belonged to Captain Spratt. It had two conical valves within the tube and a butterfly valve at the base so that it could hold a bottom water sample above the sediment core. The *Hydra* sounder was definitely an improvement because the weights drove the sampler deep into the sediment before they were thrown off (figure 56).

Shortland used a Hodges accumulator on the sounding tackle and by experiment was able to calculate the strain on the line by measuring its extension (figure 57), thus marking a further advance on Spratt's use of a spring balance. The accumulator also protected the line from breakage due to the heaving of the vessel in high seas and was essential now that soundings were being made in rough weather from large ships.

The *Bulldog* and *Hydra* sounders introduced a divergence into grabs and corers. The corer's limiting factor had always been its valve; it was essential for retaining loose material, but at the same time nothing larger than fine gravel could be collected. Scoops and grabs were effective in coarser sediment but sampled only a superficial layer, while any fine deposit was washed out during the ascent. Large stones prevented the jaws from closing; rocks damaged most sounders and could only be collected in a dredge.

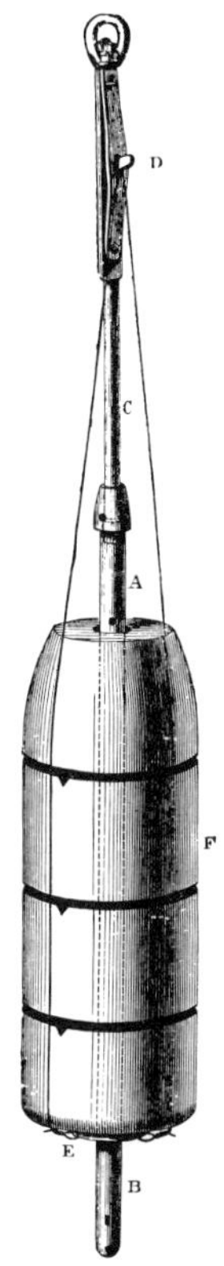

Figure 56. The *Hydra* sounder. From *Challenger Report Narrative* 1885 vol. 1, Pt 1, Fig. 13.

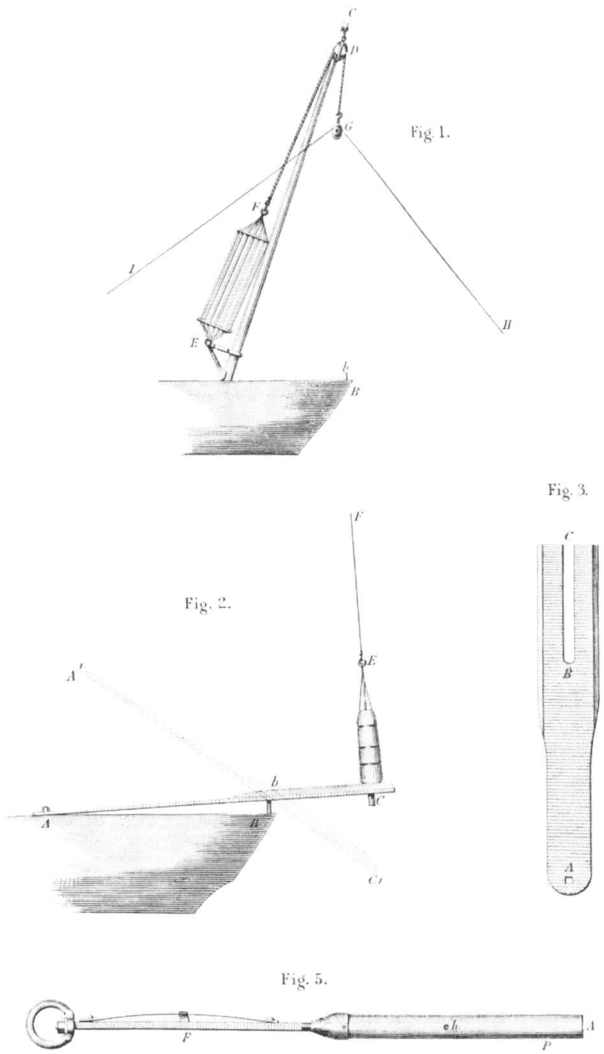

Figure 57. The *Hydra* sounder as first constructed. From Shortland 1869 *Sounding Voyage of HM Ship 'Hydra' 1868* Figs 1, 2, 3 and 5.

5.4. NEW TECHNIQUES FOR SOUNDING: WIRE LINE AND THE SOUNDING MACHINE

The *Hydra* sounders proved popular. Without the conical valves they were used by Captain Calver on HMS *Porcupine* and on HMS *Challenger* on the zoological cruises described in Chapters 7 and 9. Calver sounded to depths of three miles, reporting that it took 33 minutes to lower the sounder to this depth and 2 hours 2 minutes to haul up, using a 12 hp donkey engine.[25]

The British government benefited greatly from the installation of trans-oceanic telegraphs and for this reason were willing to contribute ships and men for route surveys. However, the rate of cable laying increased dramatically in the 1870s, financed partly by the money freed when the government nationalised the inland telegraph system; the enabling Act was passed in 1869 and came into effect in 1870. In this decade the cable companies began to build their own ships, having previously converted vessels of the appropriate capacity as required. New technologies were introduced, such as the use of wire line, the use of deck sounding machines, the invention of electrical sounders, and some novel forms of pressure sounder.

Under strong pressure from cable company shareholders, however, engineers could not afford to waste time experimenting and perfecting apparatus. Numerous devices were patented (unlike his naval predecessors, the commercial engineer was in the habit of patenting his discoveries and inventions), but were not always made. These are now of interest in showing how different principles were applied to the rapid accurate mapping of the sea bed. It was painfully clear that inadequate surveys before a cable was laid could result in long and costly grappling operations to recover and repair fractures.

As related, trials had been made during the first half of the nineteenth century using copper and iron wire for sounding, but these had not been a success because wire was manufactured in short lengths and had to be spliced for deep-sea work. Breakages often occurred at the mechanically weak points at the ends of the splices and where the line kinked due to sudden slackening. The first reported success came from Sir William Thomson (Lord Kelvin) in 1872, who had come to take an interest in deep soundings through his work on telegraphs and telegraph cables in general. From his own steam-yacht he investigated several navigational aids, the compass and several types of sounder for shallow and deep-sea use. Of a wire line he recounted:

> On the 29th of June 1872, I sounded from the *Lalla Rookh* . . . with a lead weight of 30 lbs, hung by 19 fathoms of cod line from another weight of four lbs attached to one end of a three mile coil, made up of lengths of piano wire spliced together and wound on a light wheel, about a fathom in circumference, made of tinned iron plate. The weight was allowed to run directly from the sounding wheel into the sea, and a resistance exceeding the weight in water of the length of the wire actually submerged at each instant was applied tangentially to the circumference of the wheel, by the friction of a cord wound round a groove in the circumference, and kept suitably tightened by a weight.[26]

Thomson's position at the time was inshore of a point marked 2600 fathoms in the Bay of Biscay. After running off 2000 fathoms he was wondering whether the resistance was too great, when suddenly the wheel ceased to turn. He alone among those on board was confident that bottom had been reached, but as the wire was being hauled in he realised that the winding reel was under tremendous pressure and likely to collapse, with the loss of the entire apparatus. However, despite reel distortion, he was able to recover the line with some difficulty and the attached sampler proved to be full of sediment. 2700 fathoms had been paid out and the line appeared to have run vertically down so that Thomson considered this to approximate the actual depth.

The apparatus that Thomson had constructed to pay out, measure and wind in his wire represented the first 'sounding-machine'. This term, though it had loosely applied previously to any type of sounder, was henceforth applied exclusively to such apparatus. To be used effectively, a wire line required such a machine; conversely the introduction of wire, small in bulk compared to the same length of hemp, made that machine feasible. The accumulator, the powered winch, and the line were all dismantled from their previous points of attachment on the ship and brought together within a single framework as a self-contained, powered appliance.

Thomson expressed satisfaction with his piano wire but it was supplied in 100-fathom lengths and had to be joined. The splices were about two or three feet long, soldered and served with twine, yet the wire still tended to break at the weak points at the splice ends. At his request the Manchester wire-making firm of Johnson and Nephew drew a three-mile length of crucible steel wire, in itself a considerable technical feat,[27] but the best piano wire, of high temper, could not be produced in such lengths. Since the latter gave the best results, Thomson reverted to its use, working to perfect the art of making reliable splices.

Rust also posed a major threat to wire lines. Thomson favoured storing the reels in limewater when not in use; in America, Sigsbee preferred storing the wire in oil, while others recommended using a galvanised wire. The galvanising process tended to weaken the wire, and Thomson was always against its use. Many years later, finding that the sounding wire broke 'as easily as glass', Prince Albert of Monaco substituted a nine-strand cable of 2.3 mm diameter with a breaking strain of 250 kg. He tried it during his 1894 cruise in the Gulf of Gascony and employed it regularly thereafter,[28] but elsewhere single wire of good quality gave adequate service throughout the century.

Six months after his trials in the Bay of Biscay in 1872, Thomson provided a sounding machine for HMS

Challenger, then about to leave on her long voyage. The sea trials were a failure and the machine was relegated to the hold for the remainder of the voyage, notwithstanding its remarkable success on US Coast Survey ships.

Commodore Ammen, chief of the US Bureau of Navigation, had studied Thomson's experiments and was convinced of the potential value of wire sounding. A Thomson machine was shipped on USS *Tuscarora*, a steam-powered sloop engaged during the years 1873–74 on a survey between California and Japan for a trans-Pacific cable route (figure 58). Her Captain George Belknap and his crew were able to correct several design faults on the machine, the most effective being to substitute a set of weights in place of the tackle used to tension the brake rope, a correction which Thomson himself accepted, this having been his original design. The first American machine did not incorporate an accumulator and consequently Belknap was compelled to reel in by hand as the speed of his auxiliary steam engine could not be constantly varied to allow for the ship's motion.

Belknap had originally taken a rope sounding machine of American design (figure 59) for sounding in moderate depths and for making serial temperature measurements, but he soon exchanged this for a second Thomson wire machine.[29] The conventional Brooke

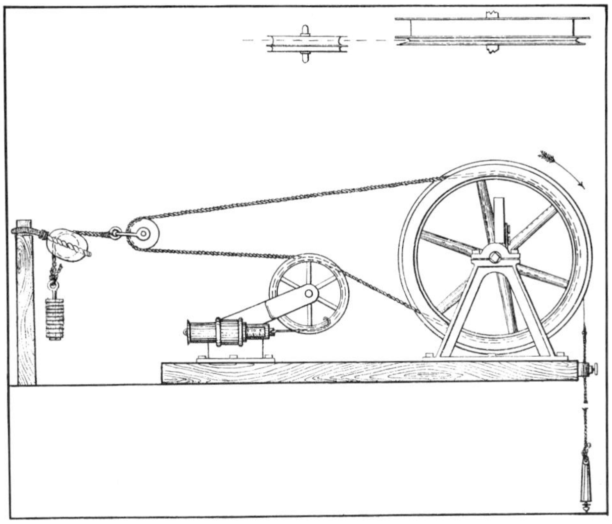

Figure 58. Layout of Thomson's sounding machine as rigged on USS *Blake* to lower a Stellwagen cup lead. From Sigsbee 1880 *Deep-Sea Sounding and Dredging* plate 6.

Figure 59. Rope sounding machine on the USS *Tuscarora*, 1873. From Belknap 1874 *Deep-sea Soundings in the North Pacific . . .* plate 19.

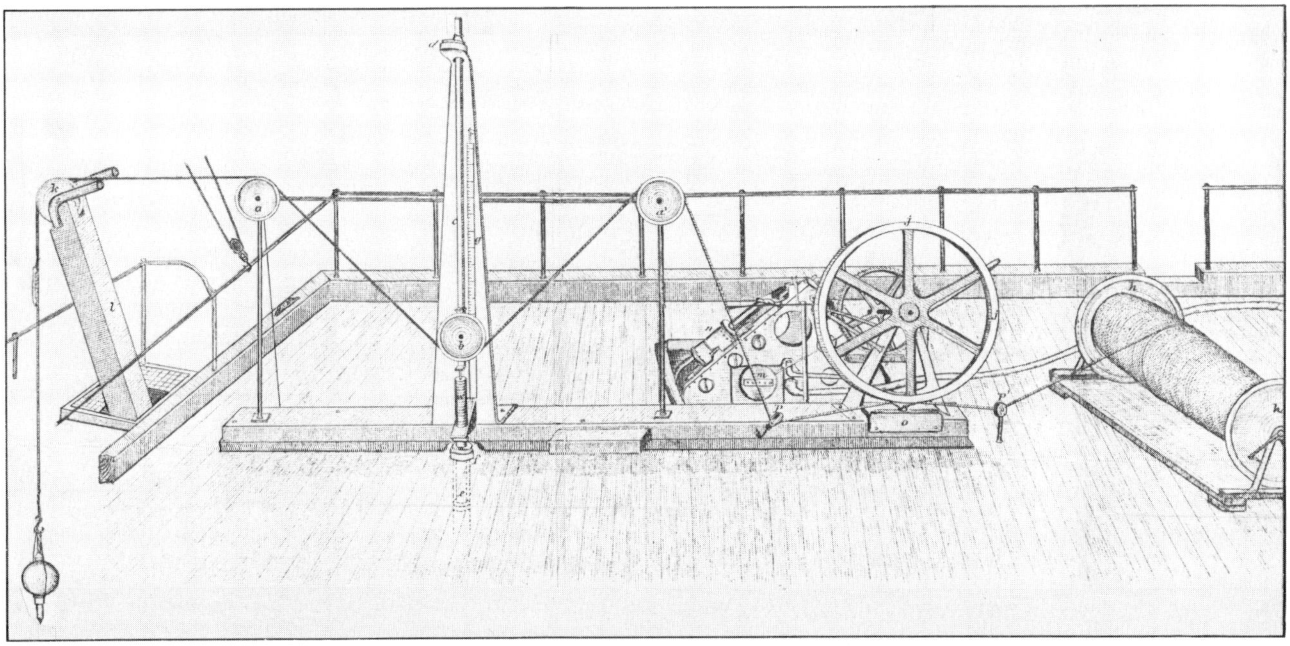

sounders were of course intended to work on a hemp line and Belknap had to adapt them to the wire. He was also given a *Lightning* sounder (which however did not impress him favourably) and a Sands deep-sea sounder.[30] Combining the latter and the Brooke sounder he devised four samplers which served him well on the various bottom sediments met with. The Brooke–Sands apparatus had an enlarged Sands cup with a modified Brooke detacher (figure 60). The lanyard and washer were retained, and the weight of the sampler cup to be hauled up was only 4½ lb.

Belknap's 'deep-sea sounding cylinder No. 1' (figure 61) collected water and mud in a valved cup, attached to a Brooke detacher. His No. 2 cylinder (figure 62) improved on this by substituting a poppet-valve for the

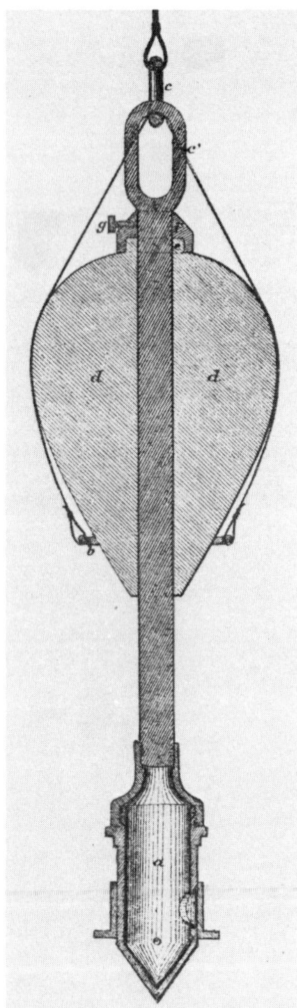

Figure 60. The Brooke–Sands sounding apparatus as first modified by Belknap. From Belknap 19874 *Deep-sea Soundings in the North Pacific* . . . plate 7.

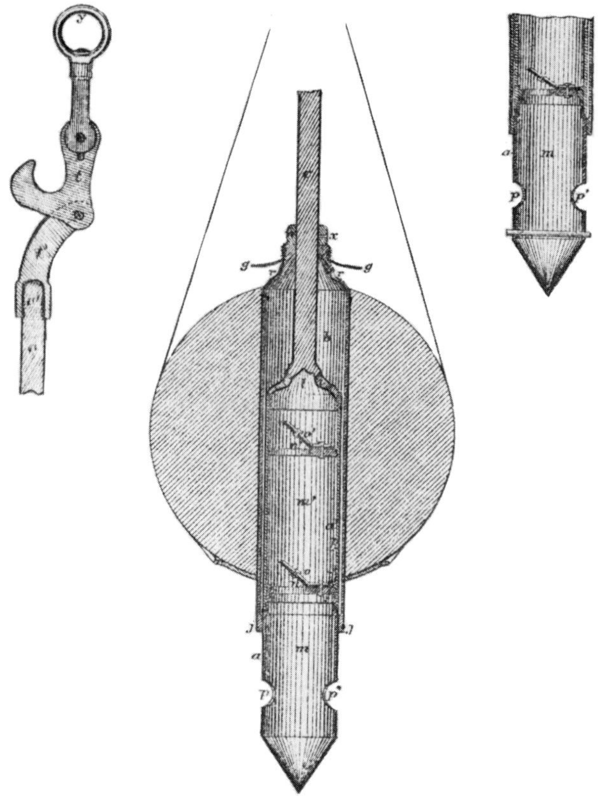

Figure 61. Belknap's deep-sea sounding cylinder No. 1 with Brooke detaching rod and sinker. From Belknap 1874 *Deep-sea Soundings in the North Pacific* . . . plate 9.

simple flaps of the first model. Cylinder No. 3 (figure 63) carried an auger-twist below and penetrated the sediment more effectively in very deep water where the apparatus did not hit the bottom with sufficient impact to drive it into the mud.

During *Tuscarora*'s survey across the Pacific 483 deep-sea casts were made, but the reel held only 5000 fathoms of wire, which was barely sufficient for the deepest soundings necessary to establish new records. A bottom sample was recovered from 4356 fathoms, while a deeper cast, to 4655 fathoms, was 'unconfirmed' in that the wire broke while hauling up and the sampler was lost. In a letter written from his ship Belknap wrote:

> I expect to accomplish all the work with the apparatus of Sir William Thomson which the Department has assigned to me, and if I succeed no greater compliment could be paid to the genius of the inventor.[31]

Sir William wrote in July to his chief at the Bureau of Navigation:

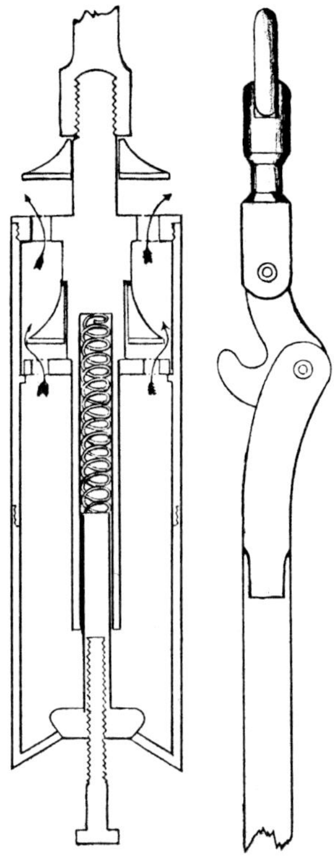

Figure 62. Belknap's deep-sea sounding cylinder No. 2. From Belknap 1874 *Deep-sea Soundings in the North Pacific* . . . plate 11.

> I have therefore resolved to alter the apparatus considerably and am preparing for trials on a new plan which I hope to carry out . . . and shall write to you reporting the result.

More than a year previously Thomson had helped Belknap and he alluded to the wisdom of Belknap's modifications and his excellent results.

Another Thomson machine had been issued to the US Coast Survey steamer *Blake*, engaged on a lengthy survey of the Gulf Stream and the Gulf of Mexico. Her commander Charles Sigsbee followed Belknap's work with interest, and during the summer of 1874 he designed and had made an accumulator with helical springs, so that the steam winch could wind in the line. That winter, appointed to command *Blake*, he built his first experimental sounding machine (figure 64).[32] The accumulator was a great improvement, and Sigsbee went on to alter the form of the machine, and to make changes in the method of work:

> The modification of improvement made by me on the original Thomson sounding machine lies chiefly in the employment of a peculiar kind of accumulator, and its adaptation to the various uses of accumulator, dynamometer, brake, correct register and governor. The accumulator eases the jerks that may be brought upon the wire while reeling in, and as a dynamometer it shows the strain upon the wire at each instant during the same operation. The brake gives a handy means of applying resistance to the reel without weights and, with the attachment of spring scales, provides a second dynamometer which during the operation of paying out shows the tension upon the wire and the resistance which the wheel suffers. The brake, in connection with the accumulator, operates as a governor on the motions of the reel when paying out wire. The odometer, used as a register in a special place, gives without interpolation the amount of wire played out.[33]

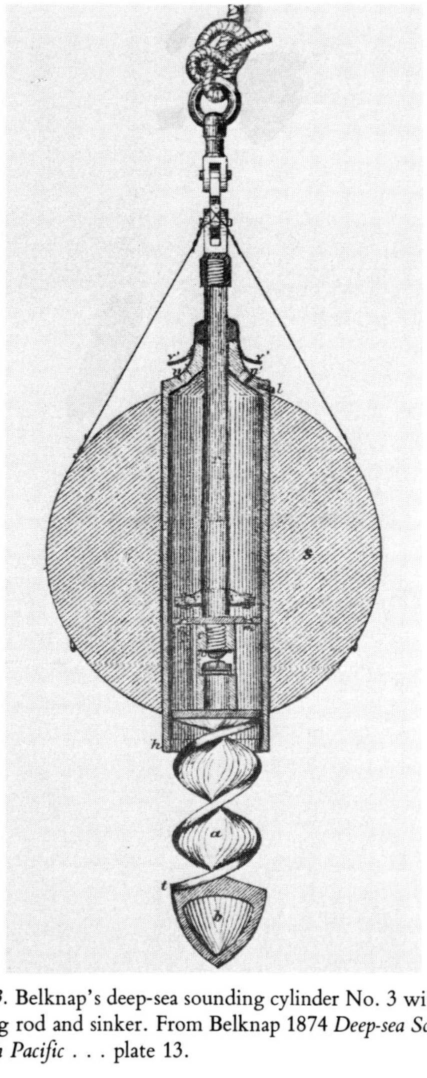

Figure 63. Belknap's deep-sea sounding cylinder No. 3 with Brooke detaching rod and sinker. From Belknap 1874 *Deep-sea Soundings in the North Pacific* . . . plate 13.

Figure 64. Sigsbee's experimental sounding machine rigged on board USS *Blake*, 1875. From Sigsbee 1880 *Deep-sea Sounding and Dredging* plate 7.

The winding-on reel was made heavy enough to prevent its being crushed by the pressure of wire, but was immediately braked by the accumulator when the sinker struck bottom. This machine was used for three years on *Blake*, alterations being made from time to time, and it gave excellent service.

The second machine which Sigsbee designed was not installed until Commander Bartlett took over *Blake* at the end of 1878 (figure 65). It carried all the components considered essential in future sounding machines; namely, a light steel reel for wire storage, a register to record revolutions of the paying-out reel, a reeling-in (or strain) pulley, an accumulator, a swivel pulley and spring scales (figure 66). On this machine the register was fixed to the cross-head pulley and the length of wire paid out was calculated from tables. The strain on the sounding wire was indicated by the degree of extension of the springs, and the whole apparatus could be easily folded up for stowage or transit. It was far more compact than the contemporary model designed by Thomson, and Sigsbee remarked with satisfaction:

> The small space required for a sounding machine by wire is hardly to be compared with that formerly taken up by sounding rope and its apparatus.[34]

This machine, and others used on board US ships, were made by Daniel Ballauf, a Washington model-maker.

Sigsbee followed Thomson's advice to insert a length of hemp between the wire and sinker, to prevent the wire from kinking if it met the sea bed. Thomson had linked the wire and hemp by an iron ring, but this would not pass over Sigsbee's pulley without slowing down the reel at the critical moment when sinker and hemp were being tossed about in the waves, with the result that the wire often broke when the sounding operation had scarcely begun. This led him to perfect the wire–hemp splice. With small weights placed on the hemp, the reel could then be run at constant speed throughout the operation (figure 67).

In the western North Atlantic, Sigsbee now found the sampler tube the least satisfactory part of his apparatus:

> While much ingenuity has been displayed in designing combined leads and specimen-cups that may be operated when it is intended to recover the sinker, there is nothing known to me that I think altogether acceptable for use with wire and on all bottoms. It should be recommended, when considering the requirements for such a device, that it would be the one most often used in depths varying from 100 to 1000 fathoms, between which depths all characters of bottom are met with—such as sand, shells, gravel, rock, clay and mud or ooze; also that retaining the sinker the whole force of impact due to the weight and velocity of the sinker at the instant of striking bottom is brought to bear on the specimen cup and unless this crushing force be provided against, the cup may be injured by a blow on a rock or other hard material.[35]

Between 1874 and 1878 Sigsbee relied on three types of sounding leads. A Stellwagen cup (figure 54) served

Figure 65. Sigsbee's improved sounding machine rigged on USS *Blake*. From Sigsbee 1880 *Deep-sea Sounding and Dredging* plate 13.

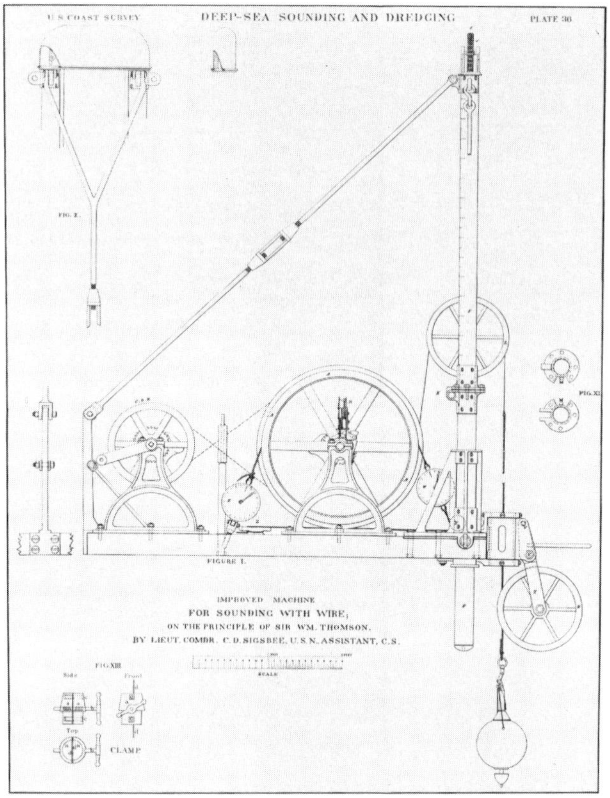

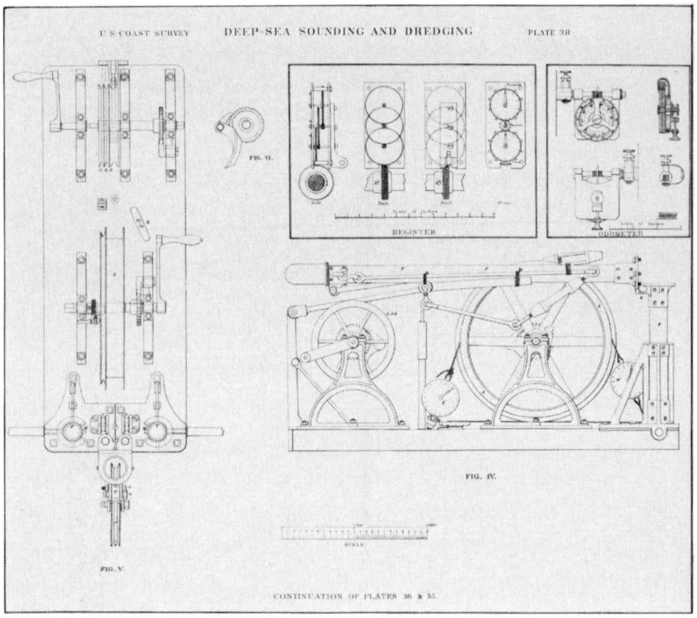

Figure 66. Technical drawings for the improved Sigsbee sounding machine. From Sigsbee 1880 *Deep-sea Sounding and Dredging* plates 36–38.

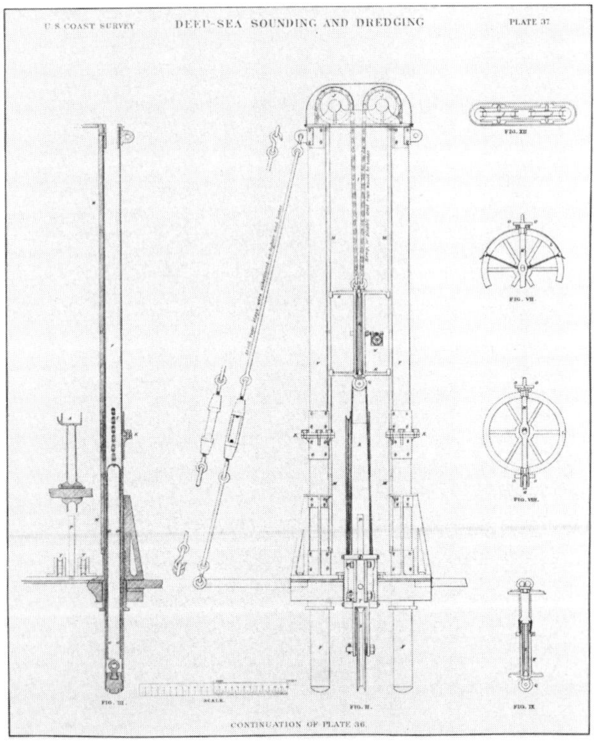

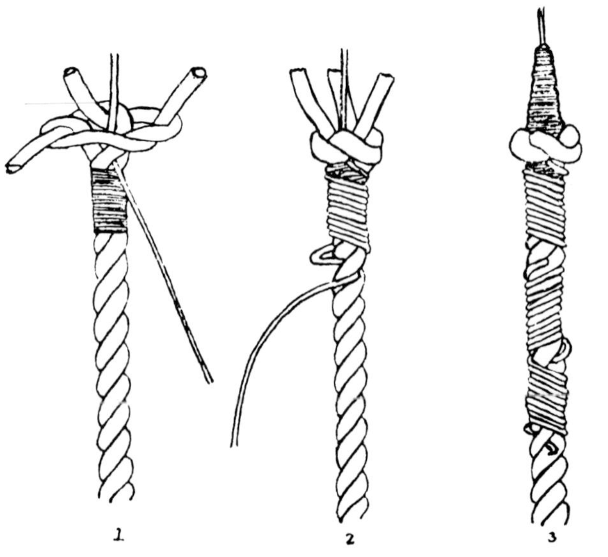

Figure 67. Sigsbee's splice. From Sigsbee 1880 *Deep-sea Sounding and Dredging* plate 41.

adequately in moderate depths from which it could be recovered.[36] For deep-water sediments the Sigsbee detacher with one of Belknap's tubes proved the best (figure 68).[37] The Brooke sounder was over-sensitive on a wire line so that Sigsbee added a stiffer detaching arm, braced by a wire spring to prevent its re-engagement when being hauled up. The collecting tube was opened at the sea bed and closed when raised, but the strength of the valve spring was critical to its success and had to be adjusted beforehand with regard to the stiffness of the sediment to be collected. For depth measurement alone, a weighted piece of gas-pipe was slung on a detacher. Its lightness made it rapid to use and, armed with tallow, it brought back enough sediment to show the general nature of the bottom.

Sigsbee never made cable surveys but his excellent sounding machine was soon adopted by British and American ships for this work. Although the structure and layout of such machines varied according to the

Figure 68. Sigsbee's sounding rod, 1896. Fig. 1, longitudinal section; Fig. 2, side view with sinker suspended by the wire bail; Fig. 3, Plan view and section of the detacher; Fig. 4, Back view of the detacher; Fig. 5, plate j and ring d; Fig. 6, enlarged view of the cone i, ring d, apertures p, and upper end of cylinder a. From *Bull. US Fish Commission* 1896 plate XVI p294.

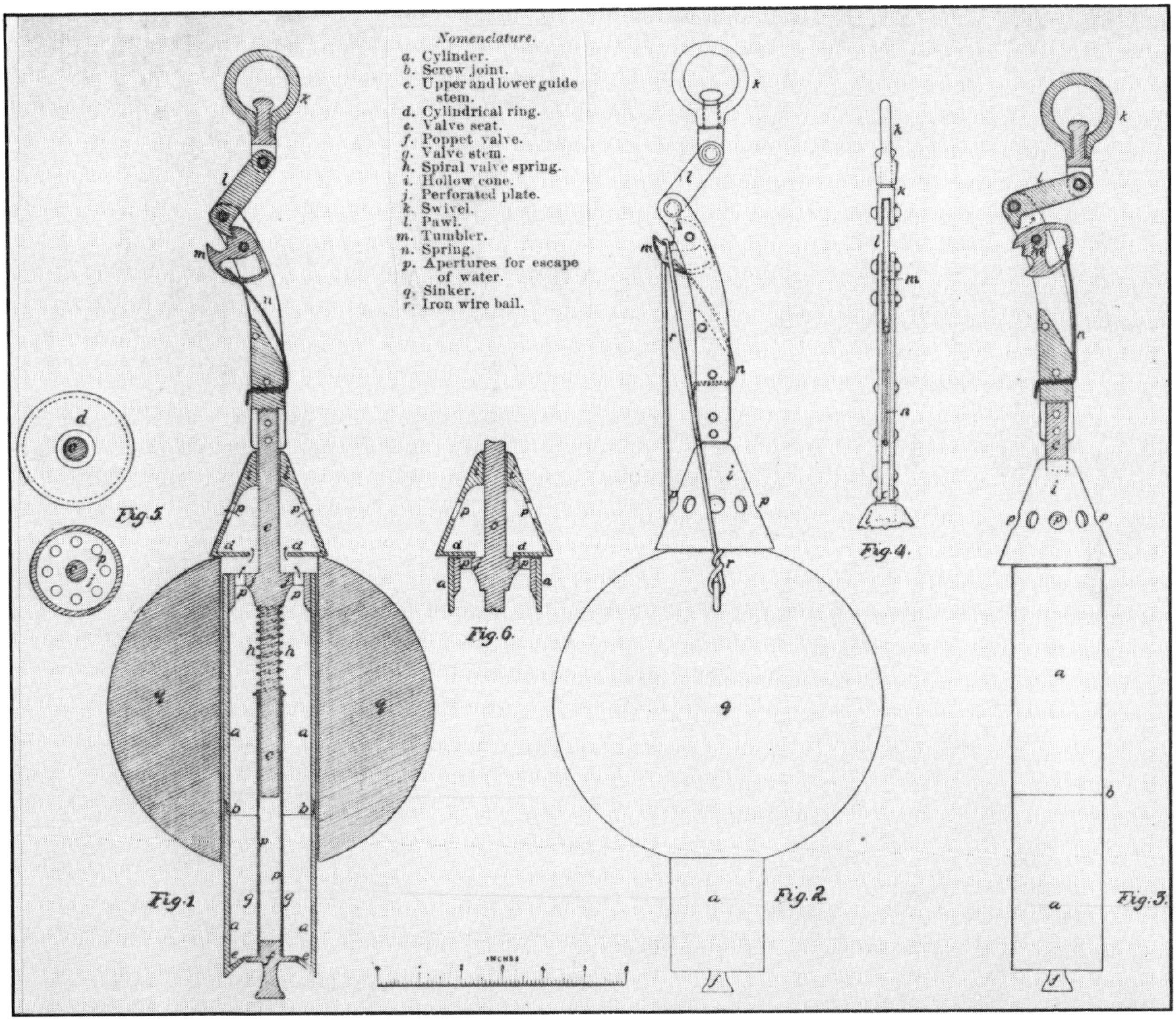

type and size of the vessel on which they served, the majority of contemporary references to deep sounding acknowledged the Sigsbee machine as the inspiration for later models. However, Thomson was never one to let a profitable idea drop, and he too worked to improve a machine for cable ships belonging to the companies in which he had a stake. The Belknap–Sigsbee samplers were not patented and probably served as patterns for similar tubes that were made by cable ship engineers between 1875 and 1880.

5.5. DEVELOPMENT OF THE THOMSON SOUNDING MACHINE

In March 1873 William Thomson described to the Institution of Engineers and Shipbuilders in Scotland a simple rope dynamometer to measure the strain put upon a sounding wire. He explained mathematically how the power of any engine could be measured with a rope weighted or attached to a spring balance.[38] He had so far made only one sounding with his machine; the rope dynamometer would undoubtedly fray after repeated soundings, due to friction against the wheel. He therefore proposed to transfer the slip to a pulley, prevented from running round by a cord whose tension was measured by spring balance, and the same rope could be used for hauling in (this had still to be done by hand, two or three men being required, and more if the depth exceeded two miles).

Thomson also considered methods of preventing corrosion of the wire. He did not favour galvanising, which weakened the metal, but preferred that the reel should be made of galvanised sheet metal. Thin slips of zinc interleaved with the wire gave full galvanic protection when the reel was stored in water. Lastly, he gave the results of tests made on various splices used to join the wire.

A year later Thomson spoke to the Philosophical Society of Glasgow about an improved sounding machine used on the cable ship *Hooper* surveying a cable route from Pará to Rio de Janeiro, sounding down to 2700 fathoms.[39] It was simpler than Sigsbee's machine, and had a totally different configuration; discarding the friction brake, he had reverted to a weighted rope as used in his 1872 machine (see figure 58). Giving his reasons for making it thus he said:

> They [the US Navy] got over the difficulty of hauling in by strengthening the wheel and detaching the weight. I prefer not to lose the lead. If the sounding is less than 1000 fathoms the wheel will not suffer.[39]

At the time that Thomson made this criticism, Belknap was however sounding in the Pacific to depths *four times* as great.

In Thomson's machine the storage wheel ran on slides along the upper frame, and was run out to project beyond the ship's rail as the sounder went down. Other pulleys came into action to assist with winding in, so that no tension was put directly upon the storage reel. It was possible to reel in as the vessel steamed ahead; indeed, flying soundings could be made in less than 200 fathoms. In this case the ship steamed at five to six knots and the wire was let go, whereupon it took up a curve between surface and sea bed which could be used to calculate the true vertical depth (figure 69).

A similar machine was built by the Silvertown Company. It too had a sliding storage reel and additional pulley systems but, unlike Thomson's machine, the reel was made of thin galvanised sheet iron. Its inertia being small, only slight friction was needed to brake it after the sinker had reached the sea bed (figure 70). If a heavy weight were put on the wire,

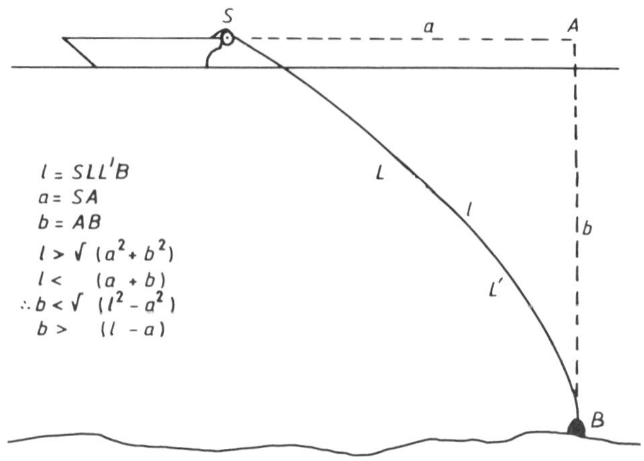

Figure 69. Thomson's formula for calculating the depths of flying soundings. From Thomson 1875 *Proc. Phil. Soc. Glasgow* **9** p116.

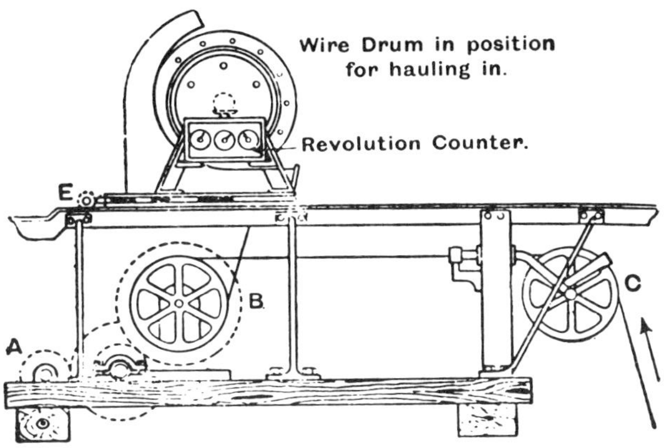

Figure 70. Silvertown sounding machine. From Wilkinson 1896 *Submarine Cable Laying and Repairing* Fig. 184.

a fast working rate was achieved. Whereas Thomson's machine lowered the sinker at 65 fathoms per minute, the Silvertown machine ran at 100 fathoms per minute.[40] With the ship hove-to a complete sounding in 2000 fathoms took only 40 minutes, but it saved time to reel in while the vessel was steaming ahead even though the recovery time was slowed by surface friction on the wire.

The French engineer Thibaudier also achieved a fast working speed with the modified Thomson machine which he put on the oceanographic ship *Talisman* in 1880 (figure 71).[41] Its sliding carriage and braking system both cushioned the wire from the ship's motion and locked the reel when the sounder hit bottom. The reel held 8000 m of wire which were reeled in by steam power.[42]

Yet another configuration was chosen by Johnson and Phillips of London (figure 72) for cable ships. Their sounding machine was self-powered, fed by steam from the main engine. The three shafts of engine, storage reel and measuring counter were connected by two endless ropes and sets of speed-reducing pulleys. With this machine the cable ship *Amber* took 380 soundings, one every seven miles, between the Gulf of Guinea and the Cape of Good Hope. The march of technical progress was as rapid in the decade 1870–80 as the operation of sounding itself.

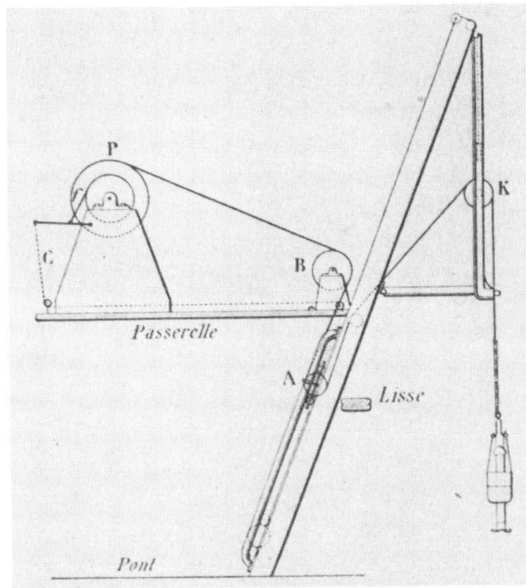

Figure 71. Talisman sounding machine. P, reel carrying 8000 m of 1 mm steel wire; B, pulley 1 m in circumference, geared to counter; A, travelling pulley linked to friction brake f by cord c; and K is a pulley designed to turn the line over the side of the ship. From Richard 1907 *L'Océanographie* Fig. 32.

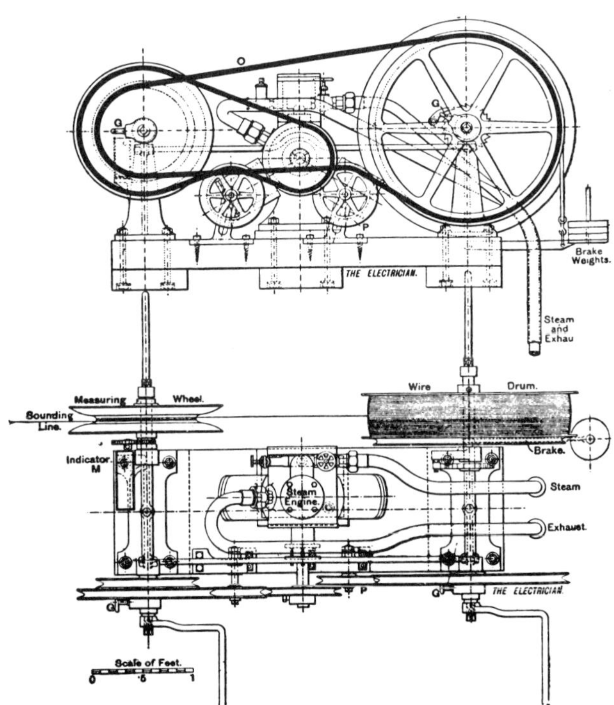

Figure 72. Johnson and Phillips' steam sounding machine. From Wilkinson 1896 *Submarine Cable Laying and Repairing* Fig. 188.

The sounding machines developed by Francis Lucas retained the necessary components within a very small, compact frame. As a youth, Lucas had served on the *Great Eastern* during the first attempt to lay a transatlantic cable in 1856, and later he became chief engineer to one of the greatest companies, Telegraph Construction and Maintenance (Telcon). His machine was given its first naval trial by Captain Nares in HMS *Alert* during 1878–79 on the recommendation of Sir George Richards, late Hydrographer to the Navy and by then managing director of Telcon.[43]

Lucas took out a provisional patent for his machine in November 1887 and enrolled the complete specification the following August.[44] His claimed innovations were:

> 2. Carrying the [paying-out] wheel within a frame which can swivel on a hollow axis the bearing for which can rock on horizontal axes and which frame is drawn upwards by a spring or equivalent and thereby caused to apply a friction brake to the reel when the weight is not hanging on the end of the wire.
>
> 3. Carrying the recording mechanism by the frame which carries the [paying-out] wheel to record the number of revolutions.

In effect, the paying-out wheel was spring-loaded and the connected friction strap round the storage reel easily regulated, initially by a threaded rod and later by a spring and hand wheel (figure 73).

This machine became standard equipment not only on Telcon ships but also on Royal Navy hydrographic vessels. Admiral Wharton, author of the standard Admiralty manual *Hydrographical Surveying*, mentioned wire sounding briefly in 1882, but his 1898 edition acknowledged that the Lucas machine had superseded the use of hemp for sounding.[45] A small machine was used for shallow soundings from boats and also for serial temperature measurements from the mother-ship. It continued to be used for nearly a century; since it also collected bottom samples it has only recently been displaced by echo-sounders and side-scan techniques. Its final form is shown in figure 74.

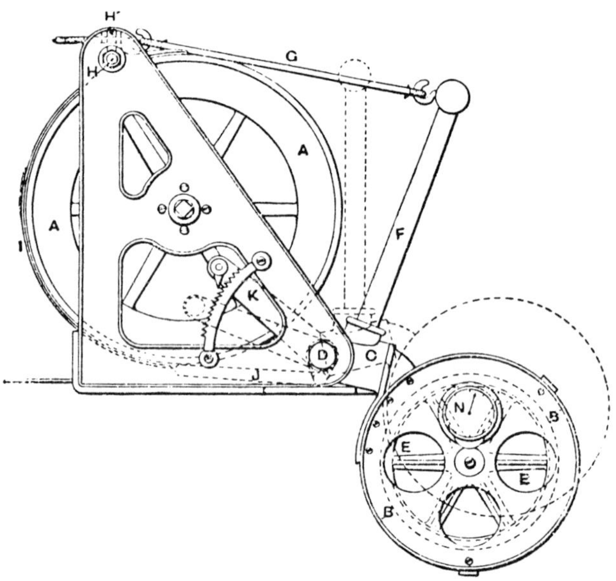

Figure 73. Lucas sounding machine, as patented, 1887. A, wire storage reel; E, paying-out reel working within frame B which can swivel horizontally around axis C, this axis being itself able to rock on pin D. Arm F, joined by spring G to lever H, enables constant slight friction to be applied to reel A. Friction band I is linked through J and C so that it is applied to reel A when B is in the upper (dotted) position. K is a locking pawl, used when preparing the machine for work (*GB Patent Specification* No. 15012).

Figure 74. Final form of the Lucas sounding machine.

Many samplers were fitted to the sounding machines described above, according to the volume of the sediment needed and the nature of the bottom. As sediment types varied little across the abyssal floor depth measurement often sufficed and simple devices like the Sherlock detacher were enough (figure 75).[46] To collect a sample either a coring tube or a grab might be put on the line. The Brooke sounder and its related forms detached their weights as soon as the lowest portion struck the sea bed. Later models were significantly improved: the weight remained attached so that the entire apparatus was driven deeply into any soft sediment, and the supporting harness was cut or slipped only as the line was hauled up. In this way a good-sized sample was taken and cores could be lengthened by increasing the weight loaded on the tube. The Silvertown corer also had an upper section to hold a sample of bottom water. Three lower tubes were fitted to hold on to soft sediments, but these could be replaced by a single tube as occasion demanded.

On gravelly bottoms, grabs such as the Lucas 'snapper' sounders were preferred, and these were manufactured in a range of sizes. In shallow water the entire apparatus was recovered, whereas in deeper water a large version with a detaching weight was used. Grapnels or dredges identified rocky outcrops. In practice each cable route presented its own problems and generated its own solutions, at least in the early decades of the era of submarine telegraphy.

Within the Atlantic and Pacific abyssal basins the Baillie sounder replaced smaller corers. This heavy apparatus was produced in 1873 (figure 116), comprising a sturdy iron tube about four feet long capable of carrying down 3–5 cwt of sinkers. Although these weights were slipped on reaching the sea bed, the tube alone was heavy enough to penetrate soft sediments to take a long core. The Baillie sounder gave good service in very deep water but was slow in use. On *Challenger*, where they worked it on a hemp line, it took well over one hour for the sounder and 3 cwt of sinkers to reach bottom in 4000 fathoms.

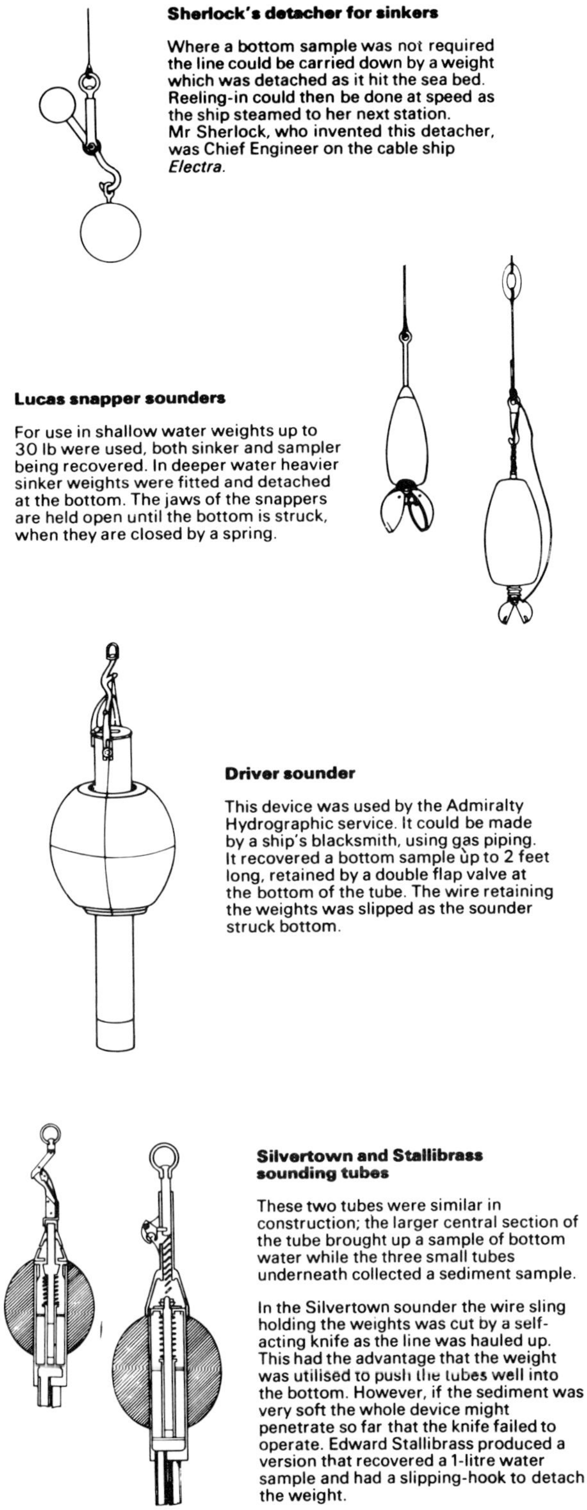

Figure 75. Cable ship corers and grabs. Redrawn from various sources.

REFERENCES AND NOTES

1. Maury M 1850 *Notice to Mariners (Explanations and Sailing Directions)* 2nd edn (Washington) pp70–1
2. Anon 1834 *Naut. Mag.* **3** p444
3. Lee S P 1854 *Report and Charts of the Cruise of the US Brig 'Dolphin'*, US Senate Document Exec. No. 59, 33rd Congress, 1st Session (Washington)
4. Faye H A E 1851 'Appareil pour sondeur à des grandes profondeurs' *C. R. Acad. Sci. Paris* **32** pp65–8
5. Brooke later filed a patent for this sounder, but the US Patent Office refused to recognise any aspect of it as an original invention.
6. Brooke J 1858 Letter reproduced in Maury M 1858 *Explanations and Sailing Directions to Accompany the Wind and Current Charts* 8th edn (Washington) pp120–1
7. Maury M 1858 *Explanations and Sailing Directions to Accompany the Wind and Current Charts* 8th edn (Washington) pp116–20
8. Maury M 1858 *Op. cit.* pp154–6
9. Dayman J 1858 *Deep Sea Soundings in the North Atlantic Ocean Between Ireland and Newfoundland made in HMS 'Cyclops' in June and July 1857* (London)
10. Anon 1844 'Instruction sur l'emploi de sondeur Lecoëntre' *Ann. Marit. et Col.*, and reprinted as a pamphlet
11. Spratt T 1856 'Bonnici's deep sea sounding instrument' *Naut. Mag.* **25** pp157–8. Spratt also summarised his thoughts on sounding in Spratt T 1865 *Travels and Researches in Crete* 2 vols (London)
12. Brooke G 1955 John Mercer Brooke, Naval Scientist, *PhD Thesis* University of North Carolina (unpublished) Ch 5
13. Spratt T 1857 *Deep Sea Soundings in the Mediterranean Sea East of Malta* (London) p9
14. Mansell A 1857 'Deep soundings between Alexandria, Rhodes and Smyrna' *Naut. Mag.* **26** pp505–12
15. Great Britain Privy Council 1861 *Report of the Joint Committee Appointed by the Lords of the Committee of the Privy Council for Trade and the Atlantic Telegraph Company to Enquire into the Construction of Submarine Telegraph Cables* (London)
16. Great Britain Privy Council 1861 *Op. cit.* pp181–6
17. Dayman J 1858 *Op. cit.*
18. McClintock L 1861 *Remarks Illustrative of the Sounding Voyage of HMS 'Bulldog' in 1860* (London)
19. *US Patent Specification* No. 2899 (1842)
20. McClintock L 1861 *Op. cit.* p11
21. Wallich G 1862 *The North Atlantic Sea Bed, Comprising a Diary of the Voyage on Board HMS 'Bulldog' in 1860* (London)
22. McClintock L 1861 *Op. cit.* p12
23. Hoskyn R 1862 *Report of the Deep Sea Soundings made in HMS 'Porcupine' in June, July and August 1862* (London) Tabulated results, p8
24. Shortland P 1869 *Sounding Voyage of HM Ship 'Hydra' 1868* (London)
25. Wyville Thomson C 1873 *The Depths of the Sea* (London) p224
26. Thomson W 1875 'On deep-sea sounding by pianoforte wire' *Proc. Phil. Soc. Glasgow* **9** pp488–93
27. Thomson W 1875 *Op. cit.* p483
28. Richard J 1900 *Les Campagnes Scientifiques de SAS le Prince Albert 1er de Monaco* (Monaco) p46
29. Belknap G E 1874 *Deep-sea Soundings in the North Pacific Obtained in the USS 'Tuscarora'* (US Hydrographic Office No. 54, Washington) p9
30. Sands B F 1857 'Description of deep-sea sounding apparatus invented by Commander B F Sands' *Rep. US Coast and Geodetic Survey 1857*, Appendix 46, p398 and Plate 70
31. Ammen D 1884 'Recent improvements in apparatus and methods of sounding ocean depths' *Proc. Am. Ass. Adv. Sci.* 33rd Mtg, pp448–9
32. Sigsbee C 1880 *Deep Sea Sounding and Dredging* (Washington)
33. Sigsbee C 1880 *Op. cit.* p60
34. Sigsbee C 1880 *Op. cit.* p67
35. Sigsbee C 1880 *Op. cit.* p52
36. Sigsbee C 1880 *Op. cit.* p51
37. Sigsbee C 1880 *Op. cit.* pp43–6
38. Thomson W 1873 'On the rope dynamometer with application to deep-sea sounding by steel wire' *Trans. Inst. Eng. Shipbuilders Scotland* **16** pp121–35
39. Thomson W 1875 'On deep-sea sounding by pianoforte wire' *Proc. Phil. Soc. Glasgow* **9** pp111–17
40. Wilkinson H 1896 *Submarine Cable Laying and Repairing* (London) pp334–6
41. Milne-Edwards A 1882 'La campagne de dragages du "Travailleur" ' *La Nature* Pt 1 p51
42. Richard J 1907 *L'Océanographie* (Paris) p31
43. Day A 1967 *The Admiralty Hydrographic Service 1795–1919* (London)
44. *GB Patent Specification* No. 15012 (1887)
45. Wharton W 1898 *Hydrographical Surveying* (London)
46. Wilkinson H 1896 *Op. cit.* p330

Chapter 6

The Fertile Imagination: Some Other Proposals for Depth Sounding

6.1. THE PRESSURE GAUGE AS DEPTH SOUNDER

Worldwide underwater exploration, particularly by cable engineers, fired the public imagination in the nineteenth century and featured in many popular American and European magazines. The second half of the century was also the era of grand international exhibitions, first in London in 1851, and then in every major industrial city. Here were displayed those ingenious devices in which their inventors and manufacturers took most pride, and their catalogues show that oceanographic instruments had a place among these. Inventors beat a path to the Patent Office doors; engineers, electricians, naval men, each with ideas to solve once and for all the problem of measuring the vertical depth from the surface of the sea to its bed far below. Some of their devices worked in moderately deep water, while others were so utterly impractical that one might suppose their progenitors had never even seen the sea, let alone ventured out into the open ocean. They do however reveal that the naval monopoly was giving way to a more general participation in marine science and for this reason they are worth describing, though few enjoyed more than fleeting success.

Two classes of sounder measured pressure as a function of depth. In the first, a chamber of air was carried down and the extent of maximum compression registered before it was hauled up, while the second incorporated some modifications to the standard engineering pressure gauge which came into use around 1860.

One of the earliest straightforward compression chamber sounders was that patented by John Ericsson in 1835 which registered by allowing sea water to force entry into a chamber, against a compressed and contracting volume of air.[1] Trapped in the chamber, the water showed the degree of maximum pressure encountered (figure 76).[2] This sounder was certainly manufactured commercially and was still popular in 1878 when Sir William Thomson considered it suitable for use on his sounding wire. It was used mainly for navigation but by altering the proportions of the chamber its range could be increased. Thomson himself designed several pressure sounders, but only three of these came into general use. The earliest of these employed a novel form of registration, namely, the change in colour of a suitable chemical when immersed in sea water. Two patents were taken out[3] and in 1876 Thomson described it thus:

> The extent to which air has been compressed is marked directly on the interior of a straight glass tube by the chemical action of seawater on a preparation of chromate of silver with which the tube is lined internally. Between the salt of the sea and the chromate of silver a double decomposition takes place. The chlorine leaves the sodium of the common salt and combines with the silver, while the chromic acid and oxygen leave the silver and combine with the sodium. Thus chloride of silver, white and insoluble, remains on the glass in place of the orange-coloured chromate of silver lining as far up as water had been forced into the tube and the chromate of sodium

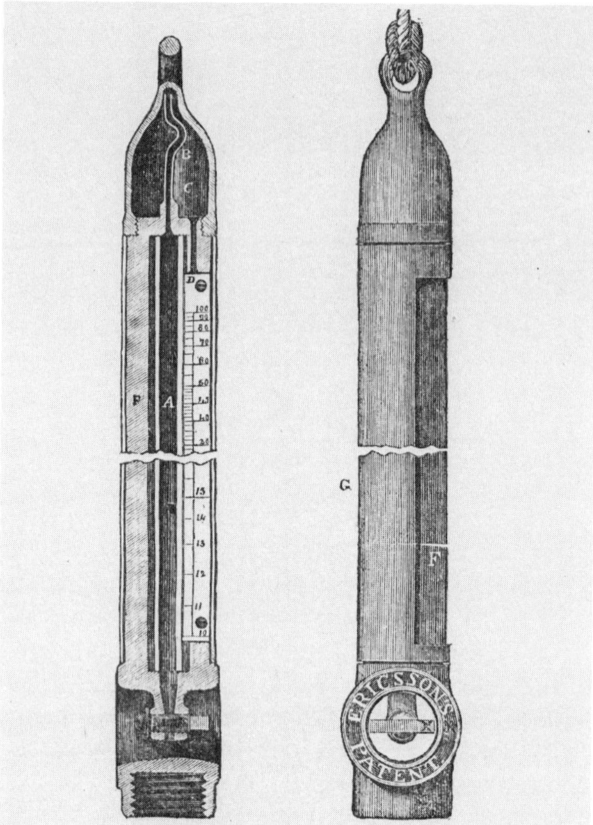

Figure 76. Ericsson's sounder, 1836. A, glass tube; B, small upper pipe. From *Mech. Mag.* **25** p353.

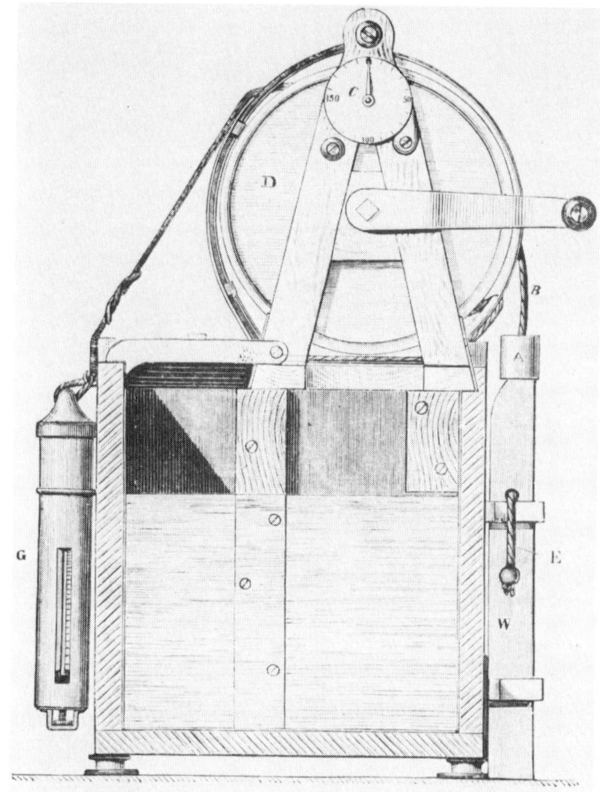

dissolved in the water expelled as the air expands when the tube is brought to the surface.[4]

He intended the pressure sounder to give a check on the wire line when making flying soundings. A tube was to be used on the first cast and from time to time thereafter to confirm that the angle of the wire through the water was correctly estimated. The glass tube, in a metal carrier, was secured to the line just above the sinker. On recovery, the chemical mark was translated into a depth measurement by laying the tube along a graduated rule.

Thomson's 1880 patent showed two other sounders. One, never commercially made, measured the volume of liquid driven by pressure out of a flexible bottle into a rigid container, a concept akin to the engineering pressure gauge. The third type shown, the so-called 'triple depth gauge', was manufactured and used for some years (figure 77).[5] The long fragile tube was replaced by three short tubes of different diameters, adjusted to measure respectively depths of 20–50 m,

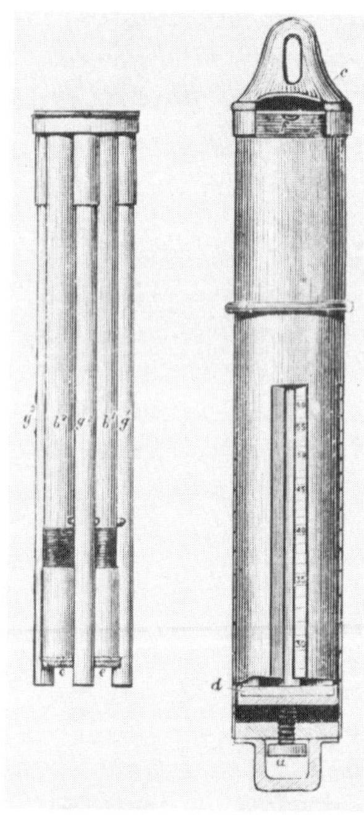

Figure 77. Thomson's triple depth gauge. From Thomson 1881 *J. United Services Inst.* **25** Figs 1–3.

50–107 m and 107–230 m. Each gauge was simply a brass tube connected by a narrow aperture to a glass tube. Fine cloth covering the end of the brass tube acted when wet as an air-tight membrane, allowing sea water to be forced into the tubes against the compressed air within. Each gauge could be read against a scale fixed on the protective container.

Yet another sounder was patented by William Thomson in 1885 (figure 78), and was also offered for sale until the early years of the twentieth century.[6] Compression drove a piston along a tube against the pull of a coiled spring, and a marker carried along by the piston remained to show the greatest pressure reached. Thomson was not the only person to consider chemical registration for this type of sounder, however; a patent by Thomas Walker in 1879 covered much the same ground.[7]

Sounders based on the engineering gauge were smaller, more sturdy instruments. Henry Johnson made three attempts to perfect a gauge in which he measured the compression of water, rather than air.[8] His first and second models (figure 79*a*, *b*) were similar. When water within the gauge was subjected to pressure an elastic ring and stopper were drawn down the stem. On being hauled up, the water expanded to return the stopper to its original position, but the ring stayed behind to show the degree of maximum compression. For his third model (figure 79*c*) Johnson replaced the moving parts by a series of glass chambers through which the water itself was forced.[9]

Figure 78. Thomson's pressure gauge sounder, 1885 (*GB Patent Specification* No. 12240).

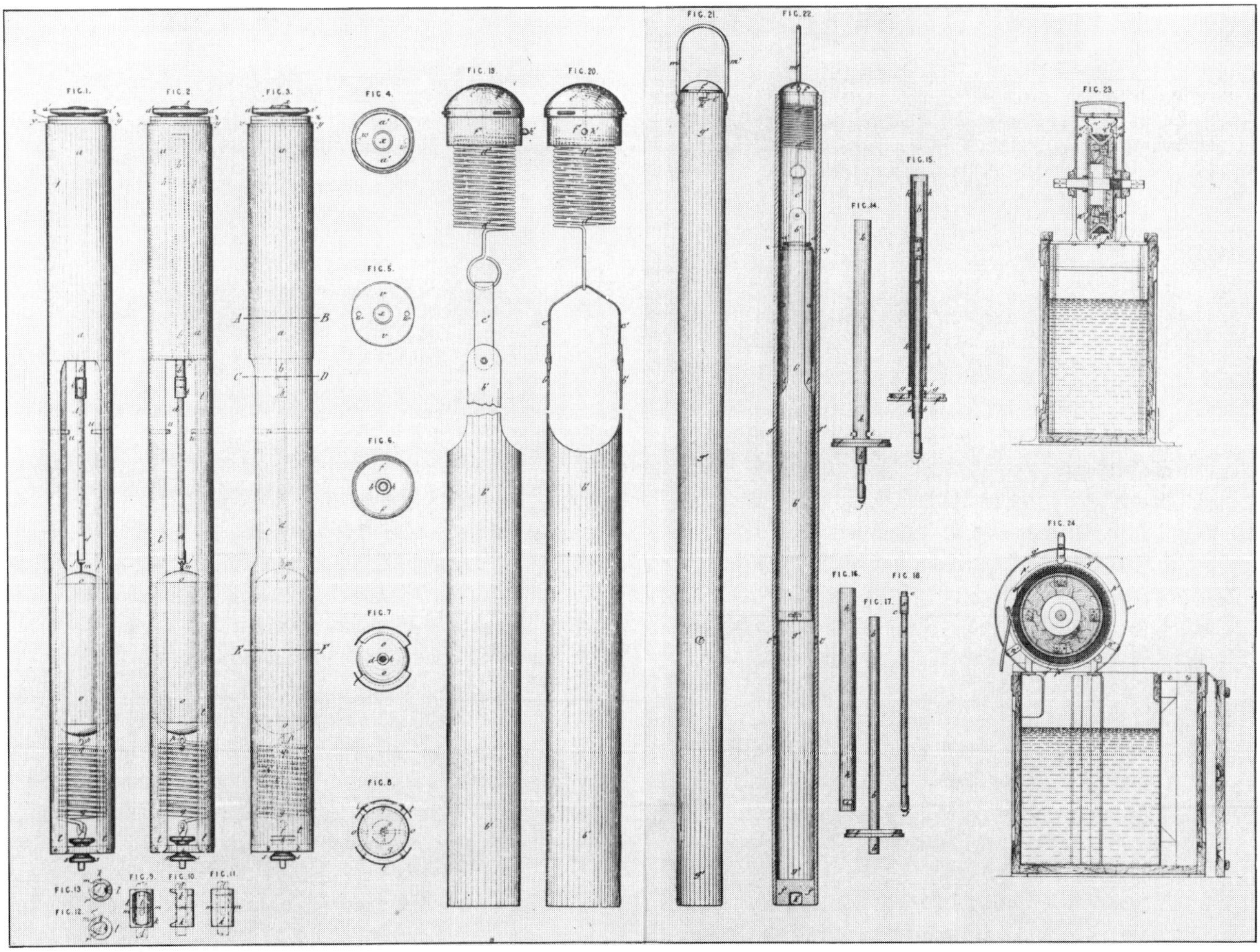

(a)

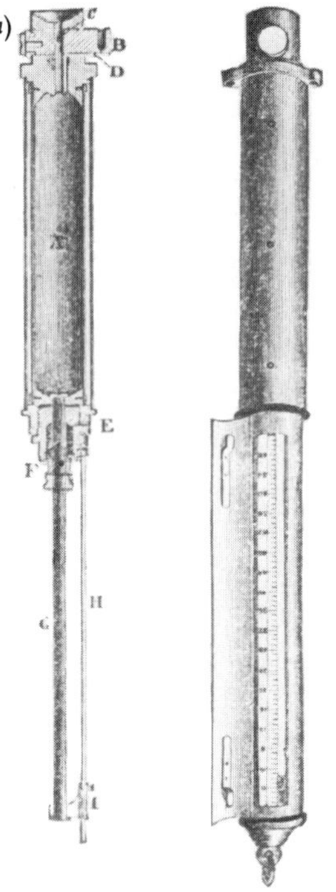

(b)

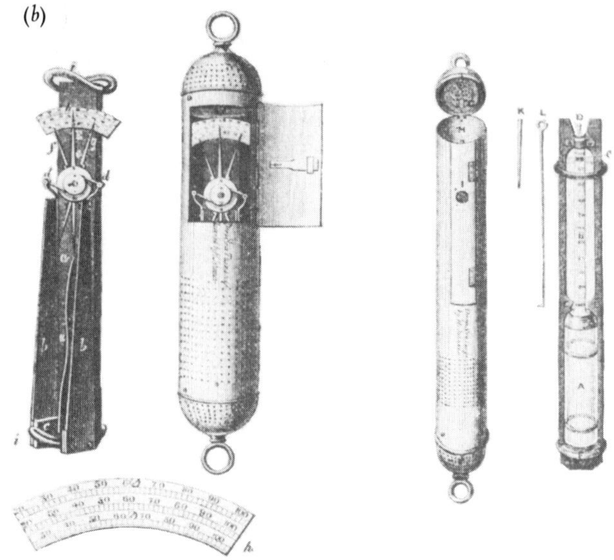

(c)

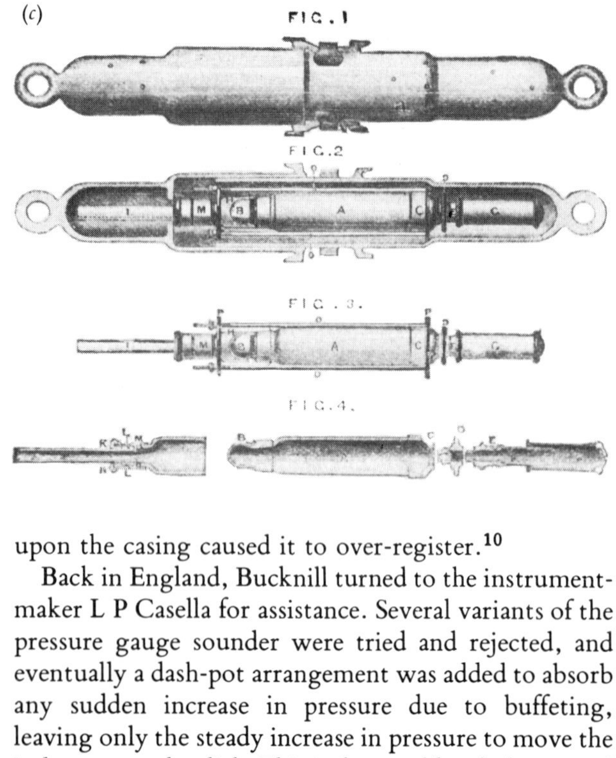

Figure 79. Johnson's pressure gauge sounders. (*a*) First model, 1869; (*b*) second model, 1861; (*c*) third model, 1866. From Glaisher *Rep. Br. Ass. Adv. Sci.* (Sections) (*a*) 1860, p203; (*b*) 1861, p58; (*c*) 1866, p26.

The first of these sounders was taken by Hoskyn on the *Porcupine* in 1862 but was found not to work. The later sounders were also tried on ships and on trials off the Isle of Wight when Johnson was working on his bimetallic thermometer (described in §8.1), but they too proved unsatisfactory.

Whilst on board one of HM ships in 1870 Lieutenant J T Bucknill of the Royal Engineers had seen the difficulties of sounding with the conventional lead line, and it occurred to him that a pressure gauge would yield more accurate results. His first design incorporated a bourdon tube gauge and dial, carried within a glass-fronted metal case. Clear glycerine filled the case so that the dial could still be read through the liquid, and the external water pressure was transmitted via an elastic diaphragm let into the top of the instrument. This prototype was constructed for him by an instrument-maker in Boston, USA. Since the pressures within and without the case were equalised there appeared to be no need for exceptionally sturdy construction, but when tested it was found that blows upon the casing caused it to over-register.[10]

Back in England, Bucknill turned to the instrument-maker L P Casella for assistance. Several variants of the pressure gauge sounder were tried and rejected, and eventually a dash-pot arrangement was added to absorb any sudden increase in pressure due to buffeting, leaving only the steady increase in pressure to move the index across the dial. This index could only be turned back by hand and so registered the maximum pressure without being returned to zero as the instrument was hauled up (figure 80).[11] The sounder was tested by Casella in the high-pressure tank he had built for thermometer trials, and also by the Navy. They found it effective and so exhibited it at the 1876 exhibition in London.[12]

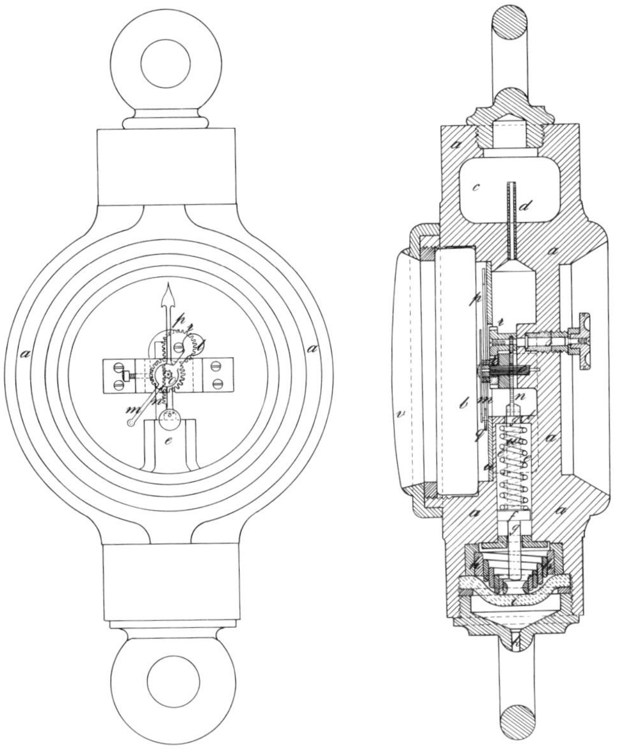

Figure 80. Bucknill's pressure gauge sounder, 1875 (*GB Patent Specification* No. 2742).

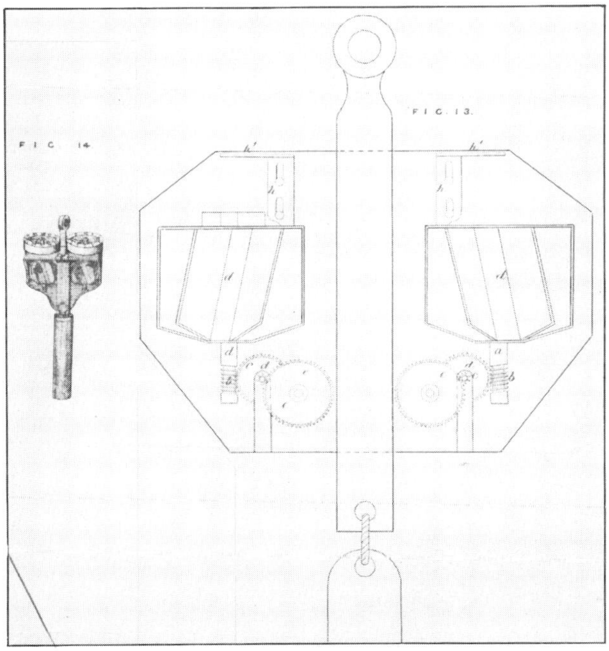

Figure 81. Patent drawing for Walker's double sounder, 1846 (*GB Patent Specification* No. 11251).

The waywiser sounder continued to be refined, and the majority were made by Massey and Walker, suppliers to the Admiralty and to foreign navies. Thomas Walker was a nephew of Edward Massey and member of a family long associated with the Hydrographic Office of the Admiralty. We have seen that not every design patented led to a practical instrument, but many of the waywisers made in the late nineteenth century still survive in museums or private collections, so there is considerable evidence to show the wide range that was made and used (see figure 82).

With a waywiser attached to one side of the lead the entire apparatus descended in a lop-sided manner, corkscrewing round the rotator axis, and the effect grew worse as chopper-bars and vane-guards were added to offset the weight of metal hanging out from the lead shaft. The error thus introduced was magnified by turbulent flow around the moving lead. It could be

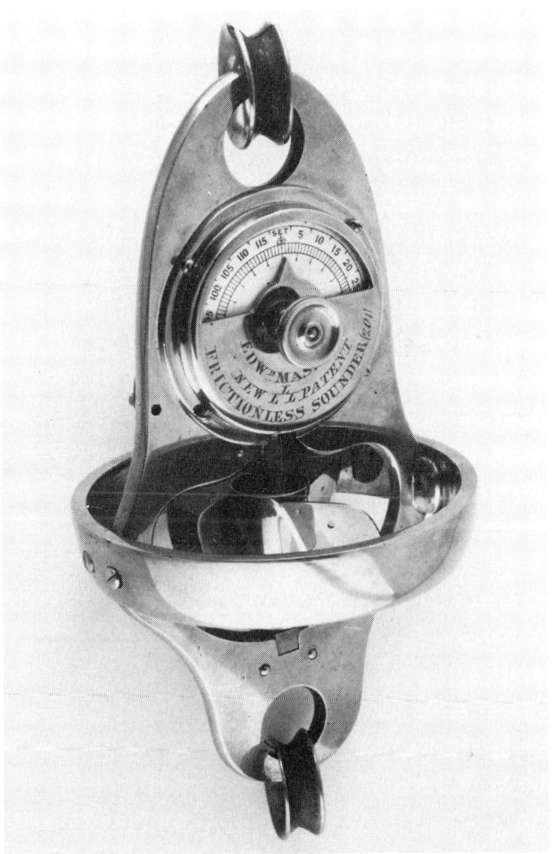

Figure 82. Massey's 'patent frictionless sounder'.

corrected by attaching a second waywiser to counterbalance the first and if its vane was pitched in reverse it would, by revolving during ascent, provide a check against its partner. Several patents were taken out for such 'dual' sounders: Thomas Walker in 1846 (figure 81); William Trowbridge of the US Coast Survey (he used two Saxton current meters as waywisers), and John Brooke of detaching sounder fame.[13]

The sounder's balance could be restored and the turbulence of its descent avoided by removing the waywiser from the lead and hanging it some distance above, let into the line on a small frame. This idea had been illustrated in Massey's patent of 1836 (see figure 30, §3.3), and was also taken up by Thomas Massey in 1857, by William Trowbridge in 1862 and by Thomas Walker in 1866.[14] The sounder shown in figure 82 was made in 1874 and may be taken as representative of this popular form at that date.

Contemporary ships' logs were often also converted into sounders. The heavy, vaned brass 'fish', towed horizontally as a log, could be dropped through the water to register depth. Many patents were filed for such instruments, but most were intended for navigation rather than for deep-sea work.[15]

6.2. SIEMENS' BATHOMETER

In 1876 the telegraph and electrical engineer William Siemens demonstrated how to register the depth of water beneath a ship's keel without putting any apparatus into the water:

> It occurred to me some years ago that the inferior quality of sea-water as compared with solid rock, such as that composing the crust of our earth, might be taken advantage of to devise a method of determining the depth of sea below a vessel. If an instrument could be constructed which, when suspended on board ship, would indicate extremely small variations in the local attraction of the earth, those indications might be referable to the depth of the sea, and a scale be obtained whose divisions would give the depth in fathoms, or other units, without having recourse to the laborious process of sounding by means of the sounding line.[16]

Some years previously his first 'bathometer' had been tried out by Dayman on HMS *Firebrand*, surveying for a cable route in the Bay of Biscay. The results were promising, but it was difficult to preserve on board the ice needed to keep the bathometer at a constant temperature.

The apparatus was in effect a gravimeter. The improved version, constructed in 1875 (figure 83), consisted of a tube of mercury delicately suspended, with a set of electrical contacts at its base. Any diminution in gravity equal to one part in 370 000, which represented about ten fathoms, should result in a proportional lowering in the height of mercury, in this case 0.00162 mm. With the addition of a micrometer screw, Siemens claimed to be able to detect changes in water depth accurate to one fathom.

Although he made calculations to take into account those factors which modify the force of gravity at the Earth's surface—temperature, atmospheric pressure, latitude and the nature of the rocks below—Siemens never got good results with this bathometer. He tried it out on *Faraday* in the North Sea and later in the North Atlantic, and obtained figures usually greater than those given simultaneously with a Thomson

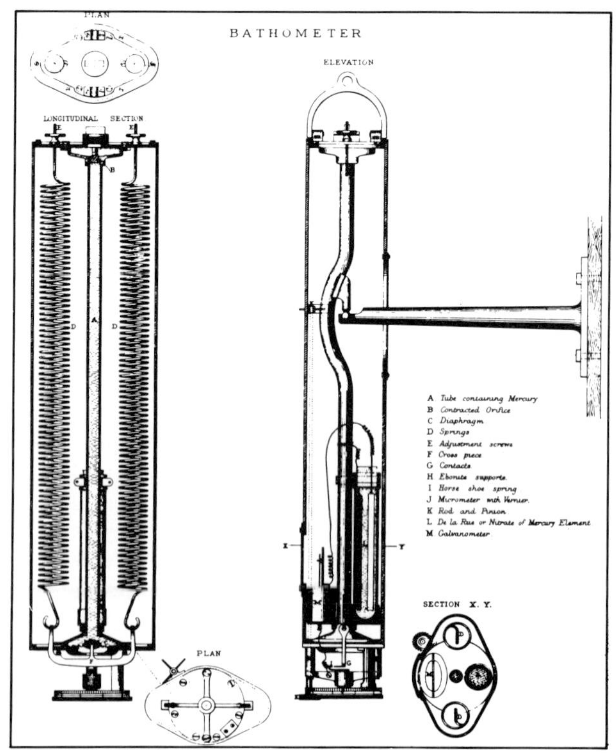

Figure 83. Siemens' bathometer, 1876. From Siemens 1876 *Phil. Trans.* **166** plate 64.

sounding machine. Try as he might, he could not account for this error but he was of course ignorant of the nature of rocks beneath the sea bed sediment and of the variations in gravity at the continental and plate margins.

REFERENCES AND NOTES

1. *GB Patent Specification* No. 6928 (1835)
2. Anon 1836 'Ericsson's patent lead or sounding instrument' *Mech. Mag.* **25** pp353–6
3. *GB Patent Specification* Nos. 3452 (1876); 781 (1880)
4. Thomson W 1878 'On compass adjustment in iron ships and on a new sounding apparatus' *J. R. United Services Inst.* **22** pp91–119
5. Thomson W 1880 'On a septum permeable to water and impermeable to air, with practical applications to a navigational depth-gauge' *Rep. Br. Ass. Adv. Sci.* pp488–9; and 'On the new navigational sounding machine and depth-gauge' *J. R. United Services Inst.* **25** pp374–81
6. *GB Patent Specification* No. 12240 (1885), and Thomson W 1891 *Popular Lectures and Addresses* **3** pp375–6
7. *GB Patent Specification* No. 3487 (1879)
8. Glaisher J 1861 'On a deep-sea pressure gauge invented by Henry Johnson Esq.' *Rep. Br. Ass. Adv. Sci.* (Sections) pp59–61
9. Glaisher J 1866 'Experiments off Ventnor with Mr Johnson's deep-sea pressure gauge' *Rep. Br. Ass. Adv. Sci.* (Sections) pp24–7
10. Bucknill J 1878 'Bucknill and Casella's patent sea-sounding apparatus for navigating purposes' *J. R. United Services Inst.* **21** pp822–4
11. *GB Patent Specification* No. 2742 (1875)
12. GB Department of Science and Art 1876 *Catalogue of the Special Loan Collection of Scientific Apparatus . . . 1876* (London) p770
13. *US Patent Specification* No. 23726 (1859); and *GB Patent Specification* No. 530 (1859) (Trowbridge); *US Patent Specification* No. 29454 (1860) (Brooke)
14. *GB Patent Specification* No. 2661 (1857) (Massey); *US Patent Specification* No. 36182 (1862); *GB Patent Specification* No. 1185 (1862) (Trowbridge); *GB Patent Specification* No. 3346 (1866) (Walker)
15. *GB Patent Specification* No. 233 (1865), for example
16. Siemens W 1876 'On determining the depth of sea without the use of the sounding line' *Phil. Trans.* **166** pp671–92
17. Great Britain Privy Council 1861 *Report of the Joint Committee appointed by the Lords of the Committee of the Privy Council for Trade and the Atlantic Telegraph Company to Enquire into the Construction of Submarine Telegraph Cables* (London)

Chapter 7

Zones of Marine Life and the Definition of Water Masses

7.1. MARINE ZOOLOGISTS MOVE INTO DEEP WATER (1830–80)

Coastal peoples have always exploited marine plants, shellfish, fishes and whales as sources of food and useful materials. Successful fishermen were familiar with the life cycles of these animals but it was in their own interests to keep such knowledge secret. When marine biology became an accepted field of study scientists took up the nets and dredges of fishermen for collecting their specimens; Marsigli himself used a common oyster dredge. In 1770 Otho Muller, a Danish zoologist, invented a naturalist's dredge having a net attached to a square frame, and this he used in Denmark's coastal waters in depths to 30 fathoms. Dredging in shallow water was so easy and its results so interesting that many naturalists turned to the study of marine life.

Deep-water dredging, on the other hand, was not so easily undertaken. The stout tow-rope made the whole assembly extremely heavy to lift, especially when the dredge collected stones or mud. A sizeable ship and crew were essential, but even so, John Ross on his second Arctic voyage of 1829–33 managed to dredge only to 70 fathoms as he depended on manpower to bring in the line, even though his ship was steam-powered. Occasionally starfish and brittle-stars were brought up on the sounding line from greater depths, and there were often arguments as to whether such creatures had truly been collected at the bottom or had attached themselves to the line in mid-water.

As deep-sea dredging was not within the capability or financial resources of most biologists, the Royal Navy was again called on to provide ships and manpower. This it was prepared to do, allocating survey ships for the task. By the mid-nineteenth century British mercantile interests had spread to every corner of the world where trading partners could be found. In this era of 'gunboat diplomacy' Britain knew that the appearance of her fleet had the magical effect of persuading reluctant foreign governments to allow trade on British terms, and a considerable number of ships were therefore kept on station around South America, in the China Seas and in the Mediterranean. They carried out surveys of ports and harbours, investigated possible cable routes and, in the eastern Mediterranean, also made detailed surveys of the Egyptian coast prior to the construction of the Suez Canal.

Survey ships were considered a naval reserve in more ways than one. With the war scare of 1852–53 the Navy had suspended construction of pure sailing ships, while recent trials showed the advantages of propeller-driven vessels. However, these advanced vessels were destined for the fighting fleet and the older sailing and paddle-driven ships were relegated to hydrographic and other 'unimportant' work. As a result, much sounding and dredging was done from paddle-steamers which meant that the gear could be handled over the stern unencumbered by a screw.

By the mid-nineteenth century naval and merchant fleets were diverging as to their construction. The Navy continued to regard steam as an auxiliary form of power, partly due to the absence of coaling stations in remote areas, while in contrast the merchant fleet, working regular passages, moved towards the exclusive use of steam, trading bunker space for smaller crews and the certainty of a faster passage through the doldrums.

Zoologists at this time enjoyed the patronage of the British Association for the Advancement of Science to a greater degree than did physical oceanographers, and so it is appropriate at this point to give an account of the origins and avowed aims of that body.[1] The Association was founded in 1831. Sir David Brewster, one of its most stalwart supporters, deplored the situation in early nineteenth-century Britain where, after the Napoleonic Wars, the country's fortunes were at a low ebb, her scientists ignored, and her industry flagging. The government, in his view, failed to appreciate the contributions which applied technology could offer to increase prosperity. Other European countries, particularly Germany, could be seen to be moving ahead, not least in the appointment of scientists to posts in government and education. The British Association was itself in part modelled upon the Deutscher Naturforscher Versammlung which had set out to bring the scientific community together and to lay its debates before government and public. It had first met in 1822; Charles Babbage and other British representatives had attended its sessions and they brought to the British Association the custom of meeting in various cities in turn, thus illuminating the needs and resources of each region of the country year by year. Like the German society, the British Association transacted its business in sections, each devoted to one aspect of science: physics, chemistry, zoology, geology, medicine, and so forth.

Brewster was the guiding spirit in setting up the Association but he wisely did so by collaborating with the flourishing Yorkshire Philosophical Society, founded ten years previously, and the first meeting was held at York in 1831. Brewster regretted the failure of the Royal Society to weld together and promote scientific effort in Britain and so he hoped that the Association would now fulfil that role, drawing upon the specialised contributions of the regional societies. Many founder members were active in more than one field of study, and among those interested in the marine sciences were Edward Forbes, Sir William Thomson, Edward Sabine, Charles Daubeny and William Scoresby. Foreign members too were not lacking; Arago, Agassiz and Bache all had connections with marine studies in their own countries. The term 'oceanography' did not appear as the title of any section; rather, this work was classed under 'geography', 'zoology', or 'chemistry'. Some 50 years were to pass before oceanography became a self-contained discipline.

Naval expeditions, as we have seen, made a point of gathering data on the physical sciences: gravity, magnetism, tides, meteorology, besides the chemistry and movements of sea water itself. Biology and geology had been served by land-based investigations, spurred on by the immense amount of construction work which revealed the nature of geological strata now exposed in railway and canal cuttings. To understand how marine life had evolved, the zoologists had to learn more about contemporary species and their habitats. Long before Charles Darwin's *Origin of Species* appeared in 1859, battle had been joined between biblical fundamentalists and those taking a more modern view of the evolution of the Earth and its various forms of life. The British Association inevitably staged many such debates, witnesses being summoned from the depths of the ocean or chiselled from the rocks to prove or disprove arguments. Marine biological work, such as that of Edward Forbes in the Mediterranean (see below) therefore received the Association's backing. The study of the ocean floor also contributed to the applied sciences, such as in the construction of offshore lighthouses and harbour works, which demanded some knowledge of sedimentation processes and submarine geology.

In 1839 Forbes encouraged the Association to set up a dredging committee and to offer small grants towards the purchase of dredges, the hiring of boats and the publication of results. At first this programme was limited in scope, although members often worked during the summer months, dredging British inshore waters. The Association issued 'dredging papers' which could be completed and returned so that some conformity of record was achieved. Gradually the work extended into foreign waters, and Forbes himself was invited to join HMS *Beacon* on her hydrographic survey in the Mediterranean, in 1841–42. Here he was able to define various zones of life, noting that plants grew only where sunlight penetrated the water. Finding that the number of animal species decreased with depth, he thought that life was probably absent below 300 fathoms in the Mediterranean although, judging from reports from Arctic and Antarctic expeditions, the boundary appeared to lie deeper in polar waters.[2]

The offshore work which continued around British coasts throughout the 1840s was undertaken largely by amateurs but some hospitality was extended by hydrographic ships and by an occasional wealthy yacht-owning patron. Gradually a concept built up, of differing zoological provinces, each characterised by a distinct faunal assemblage. Only detailed physical studies of the water made at each dredging station could explain the cause of such regional variations, and this was plainly beyond the capacity of amateur expeditions.

After Forbes' early death in 1854 the number of local dredging committees increased to the point where the Association had to set up a General Dredging Committee to supervise their activities. In 1868 the Association, together with the Royal Society, was able to convince the government of the value of a comprehensive physical and zoological cruise out into the deeper waters of the North Atlantic. The naval hydrographic ship HMS *Lightning* was put at their disposal and the scientists Charles Wyville Thomson and William Carpenter went on board to supervise the dredging operations in 1869. Together with their colleague John Gwyn Jeffreys, who joined them on a subsequent cruise, they typify those who, over the years, had contributed so much to marine zoology through the medium of the Association's dredging committees.

In 1830–32 Wyville Thomson studied natural science at Edinburgh, went on to teach in the universities of Aberdeen, Cork, Belfast and Dublin, and finally returned to Edinburgh as professor of natural history in 1870. His peregrinations gave him ample scope for dredging during the summer months, his interest in the subject having been aroused by reports of the *Bulldog* voyage of 1860. His companion, William Carpenter (1813–85), had trained in and practised medicine, but marine zoology was his spare time activity and he worked with Thomson on several collecting expeditions.

The *Lightning* cruise was considered very successful. For the following two summers the Navy provided a more seaworthy ship, HMS *Porcupine*, and expeditions were made to the Mediterranean and the Bay of Biscay as well as running over the Atlantic route covered earlier in *Lightning* in order to make more physical observations, an aspect to which less attention had been devoted during the first year. On these voyages John Gwyn Jeffreys (1801–85) also took part.

Jeffreys had practised as a solicitor and had been called to the Bar in 1856. From 1841 he had studied marine biology as a hobby and had many papers published on this topic. Later, he found himself able to devote more time to this pursuit. He superintended the general dredging committee's work, he dredged around the Shetlands during the 1860s and joined Thomson and Carpenter for the *Porcupine* cruises. Accounts of the work accomplished in *Lightning* and *Porcupine* were published from time to time both in the Association's *Reports* and in the Royal Society *Proceedings*, and when the series was over the full story was told by Wyville Thomson in his book *The Depths of the Sea*.[3] Jeffreys may have been too old to join HMS *Challenger* for her voyage of 1872–76, but he continued to go to sea, dredging in Arctic and Norwegian waters and joining the French research ship *Travailleur* in the Bay of Biscay.

British scientists were not alone in their desire to identify new marine fauna. Alexander Agassiz sailed regularly on *Blake* to collect new species off the American shores,[4] while French and Norwegian zoologists examined their own adjacent shallow and deep waters.

The progression from inshore to deep-water dredging was made possible by one logistic and two technical aids: first, to make a success of such cruises the ship provided had to be large enough to store sufficient equipment and keep all the samples collected. Secondly, steam power was essential to hold the vessel on course while towing the heavy dredge and ultimately to wind it in, a mile or more from the bottom, and containing, it was to be hoped, many creatures new to science. Only governments could make such vessels available. The third essential link, in a very literal sense, was the wire rope itself. The rope normally used for marine work was made of hemp; various qualities were manufactured, according to their intended uses, and different ways of laying or braiding, as well as the final thickness (dimensions refer to the circumference of the rope), yielded different breaking strains. The conventional British deep-sea line used for sounding was between 1½ and 3½ inches in circumference. The 3½ inch rope had a breaking strain of less than 3½ tons and was therefore inadequate for heavy dredging work.

Wire rope was developed to provide faster and safer haulage of coal in mines. Early in the 1830s a Woolwich Dockyard foreman, George Brinks, conceived the idea of a rope made up of iron wires but he was overtaken by the German Wilhelm Albert who managed to make up such a rope in 1834.[5] Brinks persevered, however, and was financed to set up a factory in Grimsby. In 1838 the Admiralty tested and approved a wire rope for rigging. This was probably not a true stranded rope but one akin to 'selvagee' made from discontinuous lengths of wire. The first patent for stranded wire rope was taken out in 1839 by Andrew Smith. Robert Newall designed a wire-rope machine which he patented in 1840 and other rope manufacturers soon entered the field. Compared with equal diameters of hemp, wire was far stronger but it retained its full weight in water (unlike the slightly buoyant hemp) and thus added considerably to the total weight put on the winches. For this reason it was not suitable for common depth-sounding operations, which continued to be made with a hemp line until this was superseded by single-wire machines.

When Brooke's sounder and others like it came into regular use on cable surveys, unfamiliar types of sediment became available for study. Examined under the microscope, deep-water muds were often seen to be composed of minute shells (or tests) of Foraminifera. A great debate then arose as to whether these tiny organisms lived on the sea bed where they were found, rather than at the surface or in mid-water, whence their empty shells fell to gather in drifts at the bottom of the sea.

The existence of deep-sea life was soon to be confirmed. In 1857 a telegraph cable had been laid in the Mediterranean between Sardinia and the North African coast and during 1860 parts of it were pulled up for repair. Some 40 miles of cable were recovered, at the Sardinian end, from waters up to 1200 fathoms. Where the cable had lain in shallow, sunlit water, it was overgrown with weed and, throughout the length examined, it had been colonised by molluscs, corals and other forms of sedentary animal life (figure 84). Undeniably these creatures had grown to maturity anchored to the cable as it lay on the sea bed.[6]

While cable engineers accepted average or approximate sea-bed temperatures, zoologists were more demanding. They looked for the small variations between

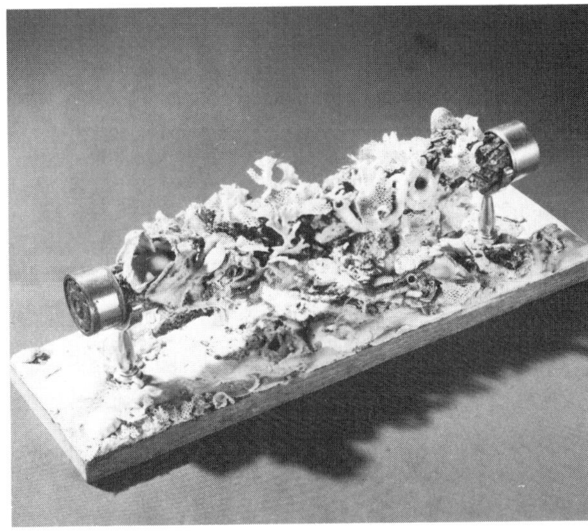

Figure 84. Portion of deep-sea cable covered with marine growths.

each water mass with its contained fauna. The thermometer had now revealed surface currents which marked shoal areas in the open ocean. Such shoals and banks usually had colder water overlying them, presumably forced up from below by submarine currents or drifts. It was remarked that this effect was not observed in the Mediterranean.

Aware that the self-registering Six thermometer did not reveal details of the temperature profile within the sea, the zoologists joined other workers who were testing other options, namely, collecting water in an insulated container, making a marine bimetallic thermometer; making a pressure-tight casing for the Six thermometer; devising a new version of the liquid-in-glass thermometer; and turning to totally different methods of measuring temperature—in this case electricity. During the second half of the nineteenth century all these possibilities were explored and tested, with varying degrees of success.

7.2. INSULATED WATERBOTTLES OF WAUCHOPE AND LENZ

The waterbottles made in the earlier part of the nineteenth century were derived from the Hooke–Hales type, with freely moving flap valves, or with flap or conical valves closed by springs—but all of these proved unreliable in use. Free valves depended for correct operation on the smooth, continuous hauling-up of the bottle. Spring-locked valves had to be triggered at a pre-selected depth and there was no way of preventing such a delicately balanced mechanism from closing too soon or too late. Moreover, the only way to insulate a bottle was by wrapping it in a slowly conducting substance, as de Saussure had done with his thermometer.

However, more experiments were carried out, and

in 1816 Robert Wauchope made a waterbottle whose insulation was partly achieved by using the water itself (this technique of putting baffles into a water mass to slow down heat exchange was not well understood at the time, but was later brought to perfection in the Pettersson bottle). Wauchope's apparatus consisted of:

> Six cases all of tin, except the outer one which was of wood, each case had a valve below and at top, both opening upwards in this manner [figure 85] the valves remaining open when descending, but closing when coming up. The four inside cases were separated from each other by a quarter of an inch all round, allowing the water to pass freely between them. The fifth case was distant from the one immediately inside it about half an inch, and the space between them was filled with tallow. The outer case of all, or sixth case, was of wood an inch thick and separated about half an inch from the fifth case by a column of water. The size of the machine altogether was about two feet high, and ten inches in diameter. To the bottom of the machine was fastened a weight of 72 lbs and to the upper part the end of a coil of two-inch rope.[7]

Figure 85. Wauchope's sketch of his insulated waterbottle. From Wauchope 1822 *Mem. Wernerian Soc. Nat. Hist.* **4** p164.

Wauchope attached a total of 1435 fathoms of rope, increasing in thickness to 3½ inches nearest the surface, with a 32-lb shot bent to the line every 200 fathoms. It took 22 minutes to veer this cumbersome assemblage overboard and it required the efforts of 100 men one hour and twenty minutes to haul in, due to the great friction between rope and water—thus accounting for the fact that such experiments were seldom made†. When recovered, a thermometer within the container was found to stand at 42 °F; the estimated depth (the line was seen to run down at an angle) was 1000 fathoms; the surface water was at 73 °F.

Between 1823 and 1826 the Russian ship *Predpriatie* sailed round the world under the command of Otto von Kotzbue. His geophysicist Emil Lenz also collected water samples for temperature and salinity measurements. On an earlier voyage aboard *Rurick* in 1815–18 Kotzbue had trouble using a Six thermometer and so on this occasion Lenz took with him two elaborate insulated containers which he referred to as improved Hales' apparatus, and which had been made under the supervision of Professor G F Parrot of St Petersburg (figure 86). Lenz subsequently described their form and function at great length, here translated and abridged.[8] All dimensions are in Paris inches (1 Paris inch = 1.066 English inch) unless otherwise stated.

> The instrument body is cylindrical, 16 inches high and 11 inches in diameter. It is made of so-called arschinenblech‡ and tinned; this material being four inches thick and 1 Arschin area. The ends of the cylinder are fitted with conical valves which are soldered to an iron rod passing through the centre. Valves and their seatings are of cast brass since rolled brass does not take the necessary polish to make the valves fit properly.
>
> The valves are not identical in size since the smaller must be able to pass through the larger seat, when the apparatus is being constructed. The lower valve is therefore 5 inches in diameter and the upper is seven inches. Small triangles brace the valve seatings to the cylinder shell.
>
> A thermometer is fixed to the central rod so that its bulb is near the lower valve, thereby lying in the coldest water. Since it was to be used down to 1000 toises§

† 1 fathom of rope 2 inches in circumference has an area of 1 sq ft, assuming that it is completely smooth. Depending on the number of strands and the closeness of lay, the actual area will exceed 1 sq ft. A thousand fathoms of such rope has therefore well over 1000 sq ft in contact with, and friction against, the water.

‡ Arschinenblech is a sheet metal named after its principal dimension, the Russian measure of area, 1 Arschin = 28 English inches square.

§ 1 toise = approximately 1.066 fathoms.

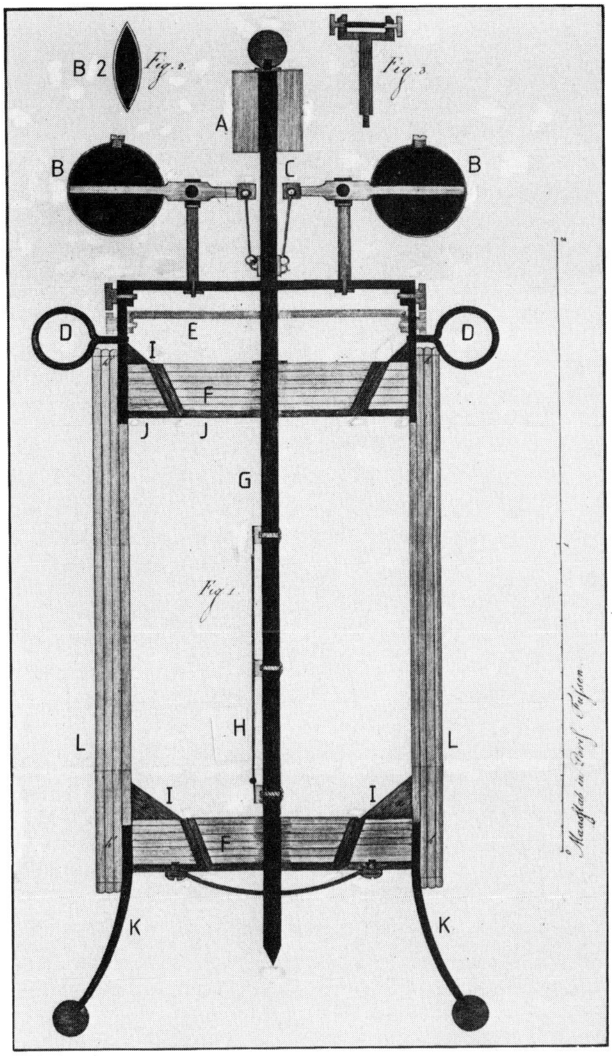

Figure 86. Parrot's water sampler, as used by Lenz, 1823–26. A, hollow vessel on the centre rod intended to balance the lead-filled bobs B,B when the sampler emerges from the water. C,C, lever system connecting bobs and centre rod. D,D, hand-holds for lifting, etc. E, cross-piece (in reality at right angles to the plane of the paper) which braces the lever frame. F,F, conical end-valves; G the rod joining them. H, a thermometer fixed to rod G. I,I,I,I, brass struts bracing the valve seatings to the cylinder walls. J,J, cast-brass valves and seatings. K,K, two of the three feet. L,L, the side wall of the cylinder showing layers of insulation (in section) which also cover valves F,F. From Lenz 1831 *Mem. Acad. Imp. Sci. St Petersburg* **1** Figs 1–3.

special thermometers of thicker glass were made. This restricted heat transfer but it was immaterial because the instrument always remained in deep water for ¼ hour and the thermometer had already been cooled on its way there. When the container was drawn up and the temperature read, this restricted heat transfer was an advantage.

Lenz feared that the free opening and closing of the valves would in some way be impaired by the pressure at 3000 toises, equivalent to 600 atmospheres, the depth that he hoped the container would reach. Ideally the valves should be weightless, but since it was not practical to counterbalance them with hollow spheres, a system of weights and levers was placed above the body of the container to ensure that they remained open on the way down and then closed until the apparatus was out of the water and safely on deck. It was this final stage of the journey that was so vulnerable, for if the counterbalance worked under the water, then it would not do so in air because of the lack of buoyancy. At this juncture Lenz calculated the thrust acting on the valve surfaces as the apparatus sank and was then hauled up. Two men could, he knew, lift it at 15 inches per second, and this rate provided adequate thrust.

Thus constructed the valves could work satisfactorily in the water but, as soon as the apparatus came out of the water the bobs would fall releasing the valves whereupon the water sample would be lost and the thermometer useless. The counterweight bobs themselves have to be balanced, but only when out of the water.

This is accomplished by placing an open vessel around the rod in such a position that it emerges from the water before the bobs do so. Its volume exceeds that of the bobs by one quarter. The water that fills this vessel has no effect upon the lever system until it comes above the sea surface, whereupon it acts to hold the valves shut against the pull of the bobs.

It remained to build up an insulation round the container. Lenz relied on Parrot's proposition that heterogeneous materials slowed down heat transfer. The cylinder and valves were covered with alternate layers of sheet iron and cloth impregnated with tallow and wax. Then the whole assembly was covered with canvas and given several coats of oil paint. As the several jackets were not in close contact, the intervening water increased the number of layers to 17 in all, so that the contained water was well insulated from the surrounding sea. With this apparatus Lenz collected water from greater depths than any samples taken by the oceanographic ships of the later nineteenth century.

Civil engineers surveying for lighthouses and harbours often took samples from shallow water. David Stevenson, member of a famous family who built many such works around the shores of Britain, designed two bottles for his own use (figure 87).[9] The shallow-water form was a simple tin cylinder loaded with lead, which went down on a line or a rod, its valve controlled by a cord from the surface. For use in deeper water, down to 40 fathoms, Stevenson made an egg-shaped vessel of lead, which went down empty and with the valves closed. A spindle projecting below the container was forced up on impact with the sea bed; this opened the valves, the bottle filled, and as it was lifted off the bottom the spindle and valves dropped again. The bottle was massively constructed and weighed about ½ cwt. Stevenson thought that it could collect water from a greater depth then he himself had occasion to work. He reported that it had served him well since 1842 and that he preferred it to the other two available types: the Marcet and *Porcupine* bottles.

Another simple valved sampler was described by Captain Denham in 1837.[10] During his survey of the Mersey estuary Denham used a 'simple gauge' to discover the quantity of silt in suspension at various depths:

> The instrument is a cylinder 17 inches long and 4 in diameter, having a valve at each end, the lower one opening inwards, the upper outwards; so that on descending both valves would be opened by the pressure of the water, which would flow freely through the cylinder, and, on ascending, both valves would be closed and the water which had entered at the lowest point of its descent be retained within it.

No illustration was shown, but this appears to be the forerunner of the standard Admiralty bottle as later taken on HMS *Porcupine* (see figure 94) and therefore probably that to which Stevenson was referring in the previous paragraph.

The waterbottles so far described have consisted of valved containers through which water flowed as they were lowered. There was always a risk that water from one level might be carried down and mixed with the sample when the valves were closed, yet if the bottle was sent down closed but empty then it was likely to be crushed by pressure during descent. An ingenious way of avoiding this dilemma was described by Johan Wilke whose experiments in Landscrona harbour were presented to the Swedish Academy in 1771.[11] Wilke's 'water extractor' was, he said,

> nothing more than a type of extractor or spray, consisting of a tin cylinder about 15 inches long and three inches wide, attached to which is a somewhat narrower and longer completely enclosed cylinder which acts as a piston and which is surrounded towards the bottom by two wide lead rings which easily fit the cavity of the outer cylinder and leave space for bound tow or leather to block the spray sufficiently slowly and serve as a pump-piston.

The original bottle was made of well-tinned iron. Wilke sent it down to 20 fathoms, and weighed the recovered samples against pure rain and snow meltwater at room temperature. His results, and the illustration of the bottle, were reproduced in the Academy's proceedings.

When Jean Baptiste Biot devised a waterbottle for the cruise of *La Bonite* (1836–39), he wanted a bottle which would not carry down any trapped air as he wanted to investigate the gases present in sea water. Whether or not he was familiar with Wilke's design, Biot constructed one of similar type: a strong glass vessel with a piston (figure 88).[12] Lines ran from fore

(*a*) (*b*)

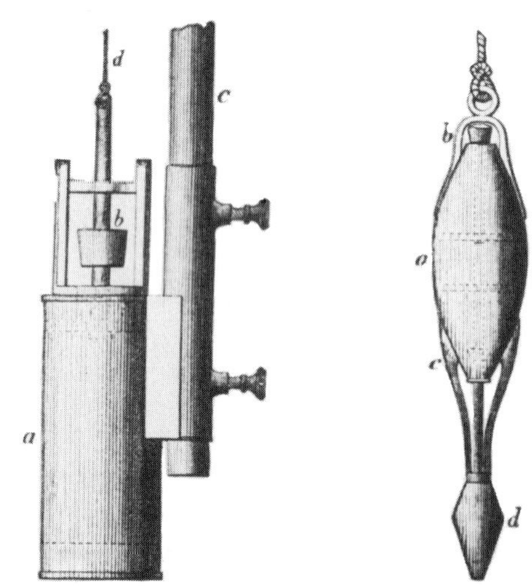

Figure 87. Stevenson's waterbottle. (*a*) Shallow-water form: a is a tin cylinder, b is a conical valve, c is a graduated rod to mark depth, and d is the cord operating the valve b. (*b*) Deep-water form: a is a leaden container, b and c are valves, d is a lead weight on the end of a spindle attached to valve c. From Stevenson 1872 *Canal and River Engineering* Figs 17 and 18.

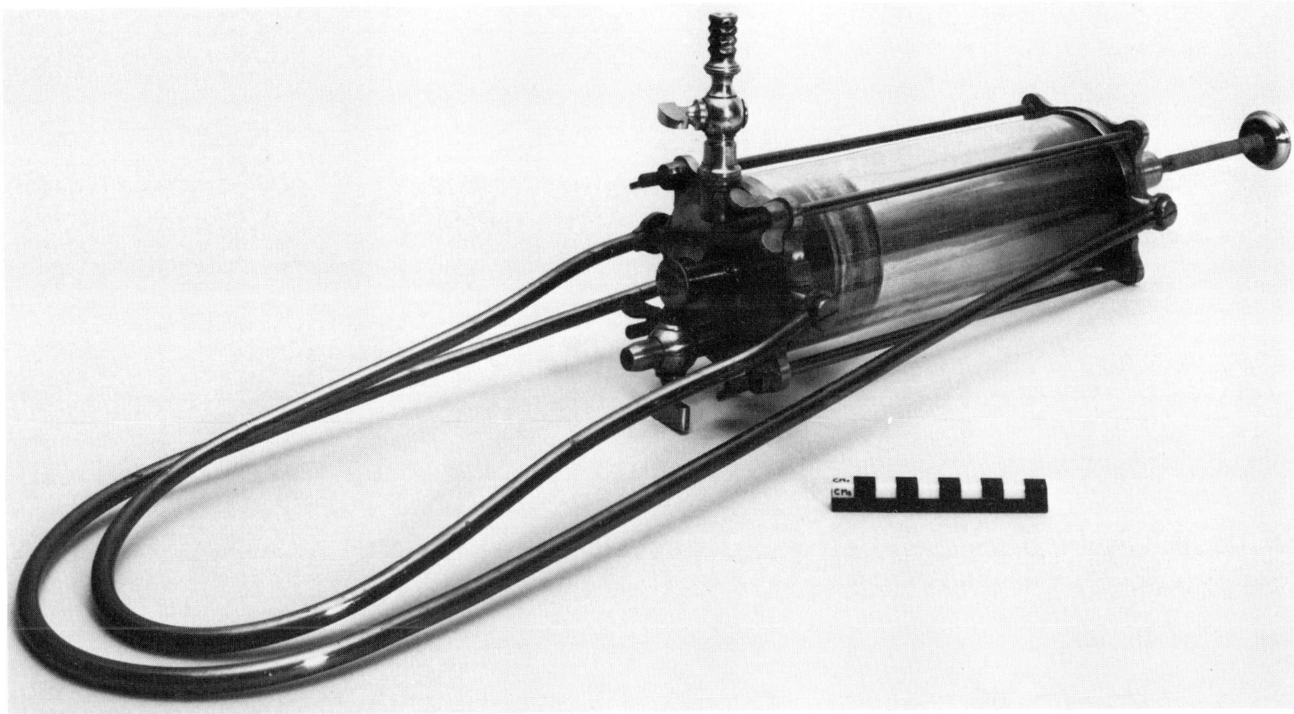

Figure 88. Biot's plunger waterbottle.

and aft of the ship to brass yokes at each end of the bottle. In this way, without the lines becoming entangled, the bottle was lowered to the required depth, and then, by hauling on the other line, reversed to take the weight on the piston and in this manner filling the bottle. As the piston was drawn up the cylinder, water entered through a small flap valve. During ascent, the container was completely water- and gas-tight.

At the same period Professor Charles Daubeny of Oxford studied the chemistry of sea water and its contained gases, and he decided to use a bottle similar to that of Biot. Daubeny exhibited his bottle to the British Association in 1836 but did not publish a description until two years later,[13] by which time he had tested it in 50 feet of water off the Kent coast. He intended to use the bottle in deeper water and claimed that its simplicity made it superior to that of Biot, and therefore worthy of consideration since all previous apparatus had been faulty to a greater or lesser degree.

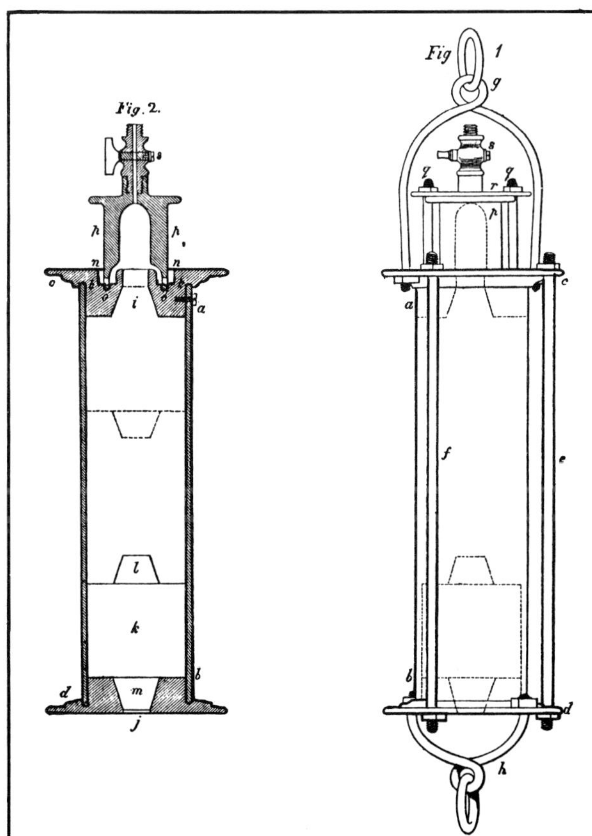

Figure 89. Daubeny's waterbottle. From Daubeny 1838 *Trans. Soc. Arts* **51** p8, Figs 1 and 2.

Whereas Biot had taken a glass cylinder as the body of his apparatus, Daubeny preferred brass, cleaning it frequently to avoid contaminating the sample. Its gas container formed an integral part of the body, unlike that of Biot, while the valves were also different. In fact Daubeny ended up with a bottle which did not merit the acknowledgment that he gave to the French scientist (figure 89).

A similar reversing action bottle was designed by Captain Rung of the Danish Meteorological Service and described by him in 1886.[14] A single line carried the bottle, looped to a detaching hook. A messenger sent down the line released the hook, allowing the bottle to fall on a captive rope, with the piston taking the weight of the apparatus. As it withdrew, the bottle filled (figure 90). The weak point of all these piston-filled bottles was the entry valve through which water might force entry during ascent, allowing the sample to be taken from more than one level.

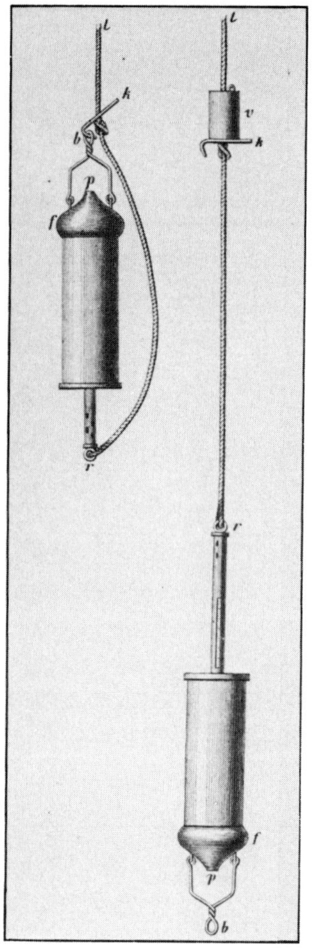

Figure 90. Rung's plunger-operated reversing waterbottle and thermometer. From Rung 1886 *Meteorol. Z.* **3** p549.

7.4. THE STANDARD ADMIRALTY WATERBOTTLE

In 1868 the Royal Society laid before the Admiralty a statement of the advantages to be gained from a short dredging cruise in the North Atlantic, with a request that a ship be made available. That autumn Professor Wyville Thomson and William Carpenter were able to sail in HMS *Lightning* and a full account of this eventful voyage was given by Wyville Thomson a few years later.[15]

HMS *Lightning* was the oldest paddle-steamer in the Royal Navy, and the bad weather encountered during the six-week cruise made conditions very hard for all on board the leaky old vessel. She sailed from Pembroke via the Hebrides to the Faeroe Islands, returning via a detour out into the Atlantic. The results showed beyond question that life in deep water was abundant and varied, represented by all the invertebrate groups down to 650 fathoms, the deepest level sampled by the dredge. Another important fact emerged: instead of deep water having a uniform temperature of 4 °C, great masses of water of differing temperatures, each characterised by a distinct fauna, were found to be in slow movement, contributing to a grand system of oceanic circulation. A large proportion of the deep-sea species captured were hitherto unknown, while others appeared identical to fossil forms previously known only from the chalk and older rocks on land which had originally been laid down on the sea bed. Small sediment samples were collected in the *Lightning* sounder scoop (figure 91).

So much of value resulted from the hauls made by *Lightning*'s dredge that HMS *Porcupine* continued the series during the summers of 1869 and 1870. *Porcupine*'s chartroom was fitted out as a laboratory and a chemist joined the two naturalists on board, so that analyses of the sea water could be made, in addition to temperature measurements.

HMS *Porcupine* was a far more seaworthy craft than *Lightning* and, under the direction of Captain Calver, her crew soon became adept at handling the heavy dredge (figure 92). In 1869 three cruises were made, each with its own scientific team. The first covered an area west of Ireland, extending as far north as the Rockall Bank. The dredge was sent to 1470 fathoms, again bringing up a varied haul of invertebrates. This led the scientists to direct the next cruise into the Bay of Biscay, the nearest area where deep water might be found. Two hauls were obtained from 2000 fathoms, each operation taking eight hours. The heaviest of the dredge frames was used, weighing 225 lb and 4½ feet across. The frame carried an outer bag of strong twine netting and an inner fine-wove lining. At Calver's suggestion, a transverse bar was fitted at the rear of the bag and on this were hung several hemp swabs of the type which the sailors used to swab the decks. These entangled some small creatures which, half buried in mud or wedged between stones, might otherwise have been missed as the dredge passed over them. A weak link fitted on the frame saved the tow-rope breaking if the dredge became fast in the rocks. As the link snapped, the frame jolted sideways and this normally freed it. A derrick was rigged, sometimes at the bow, sometimes at the stern (figure 93), with an accumulator to ease the strain on the line brought about by the ship's motion. After moving the dredge across the sea bed for several hours, the donkey engine took a further four hours to bring the gear in as the total weight of some 3000 fathoms of line, which was weighted by extra sinkers, plus the dredge, was 2042 lb.

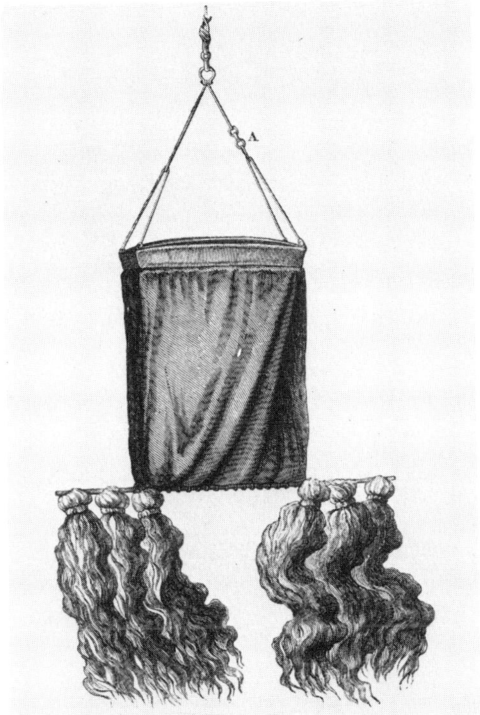

Figure 92. The *Porcupine* dredge, with 'hempen tangles'.

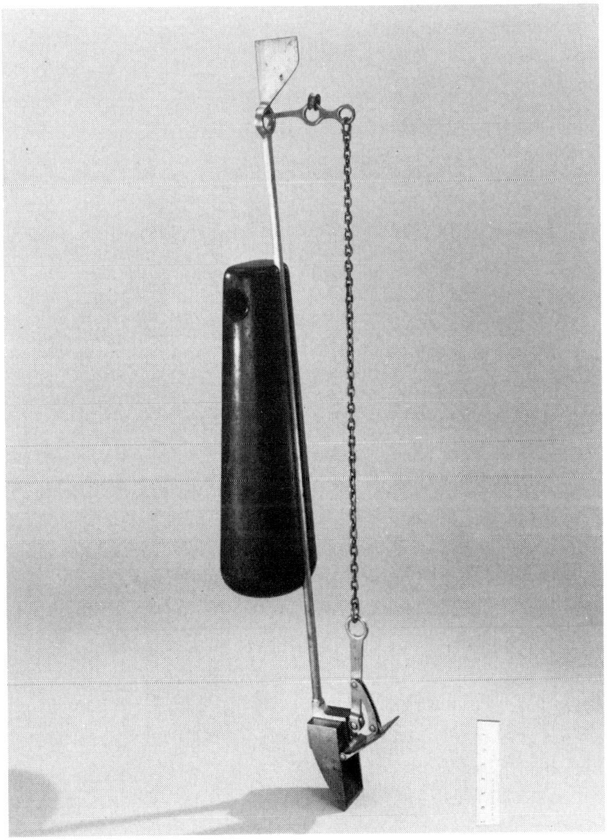

Figure 91. The *Lightning* or Fitzgerald sounder.

Figure 93. The stern derrick of HMS *Porcupine* showing the accumulator rigged for dredging and sounding gear. From Wyville Thomson 1873 *The Depths of the Sea* Fig. 46.

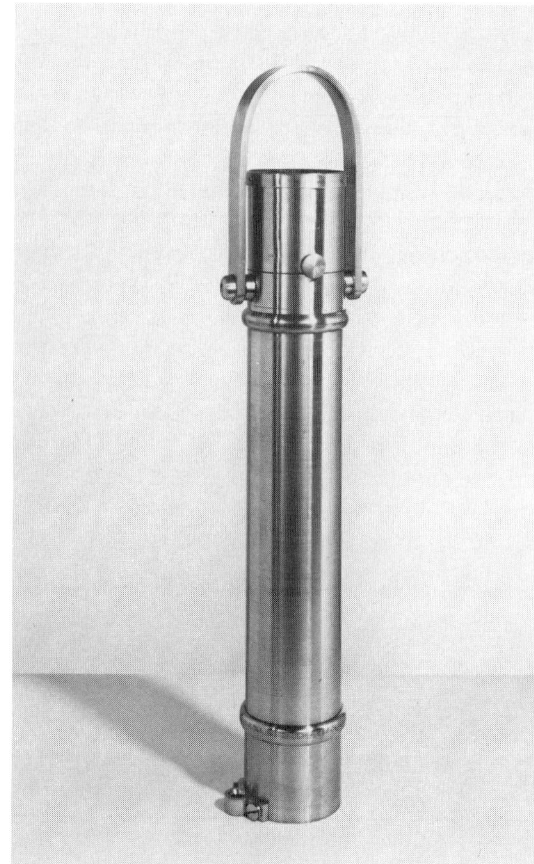

Figure 94. Standard Admiralty waterbottle, as used on HMS *Porcupine.*

For her third cruise, *Porcupine* returned to the stations covered by the *Lightning* cruise, in order to make the sea water analyses that had not been part of the routine that year. A large-capacity waterbottle was used for this work (figure 94)—a brass tube with conical valves. The inside of the bottle was coated with sealing wax to prevent corrosion and contamination of the sample. As with others of this type, the bottle worked well as long as it could be hauled up smoothly, but this was not always possible. Wyville Thomson reported:

> A perfectly satisfactory waterbottle is still a disideratum, but I believe that one that was used by Dr Meyer and Dr Jacobsen in the German North Sea Expedition of the past summer, goes far to remedy these defects.[16]

7.5. SIGSBEE'S PROPELLER-CLOSED WATERBOTTLE

Charles Sigsbee of the US Coast Survey, whose sounding apparatus was described in §5.4, also designed a small sampler to collect water for salinity and chemical analysis. It was to be a precision instrument, closing securely at depth to prevent any subsequent contamination from upper levels. Sigsbee invented this 'water-specimen cup', as he termed it, in 1875 and made several improvements up to August

1877, at which date it was drawn to illustrate his book *Deep-sea Sounding and Dredging* (figure 95).[17]

Sigsbee's waterbottle was constructed of brass throughout, except for certain bearings which were of german silver, an alloy that is not tarnished by sea water. The cylinder was closed by poppet-valves and above it was a strong frame holding a propeller. The bottle went down with the valves open and the propeller free-wheeling. When the line was hauled in the propeller was turned by water flow and it screwed down upon the upper valve, closing it and, through a central stem, the lower valve also. As the valves closed fully the propeller came to the end of its travel and slipped off the driven worm to a lower free-wheel position. In this way whatever the subsequent direction of thrust upon its blades, the propeller had no effect upon the valves which then remained securely closed. In use, Sigsbee's bottles were clamped to the line by phosphor–bronze spring clips. They served in offshore waters of the United States, but not in great depths where the mechanism might well have proved troublesome. The bottle was also used on the European expeditions of *Valdivia*, *Pola* and *Ingolf* during the 1890s. Since it required a lift of some 25 fathoms before the valves were securely closed, it was not suitable in waters where the distinct strata were closely spaced.

As this bottle was not pressure-tight, the sample lost all free gas during ascent, and was therefore not

Figure 95. Sigsbee's waterbottle. Section drawing and method of attaching it to the line. From Sigsbee 1880 *Deep-sea Sounding and Dredging* plate 40.

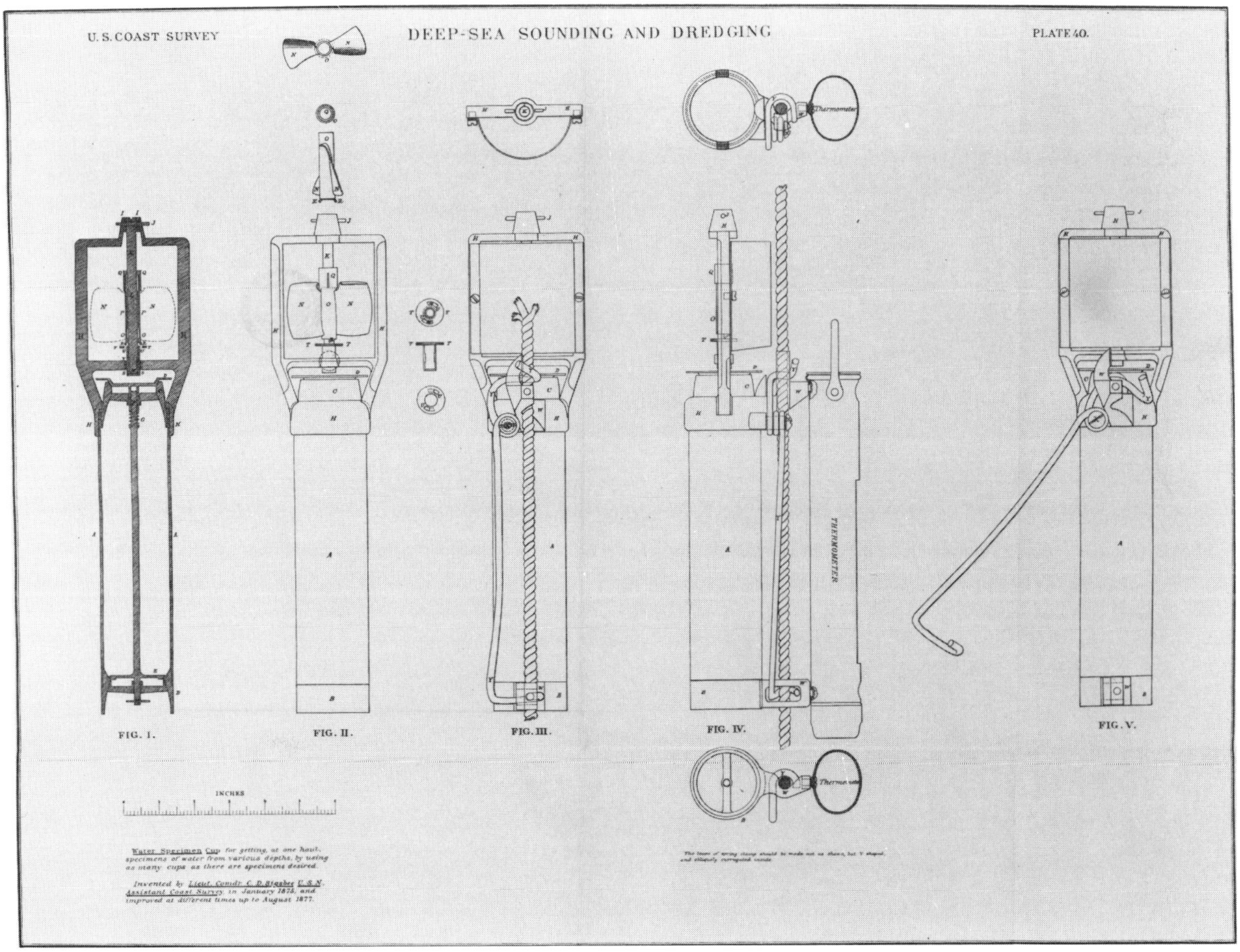

acceptable for a full chemical analysis. A much stronger bottle, known as the Kidder–Flint bottle, was designed in 1884 for the US Fish Commission steamer *Albatross*.[18] The cylinder and expansion chamber of the Kidder–Flint bottle could withstand internal pressures up to 150 lb/in^2. As the valves were no longer joined by a central stem, it was necessary to fit a propeller at either end. Like its predecessor, this bottle was used on American ships but still proved troublesome on occasion.

So ends the sequence of simple valved samplers first illustrated over 200 years earlier by Robert Hooke. They were never entirely satisfactory; those with end-valves closed by water thrust alone were likely to re-open when hauled in roughly, while mechanical valve locks were little better since bottles such as those of Scoresby, Lenz and Sigsbee still had to be hauled up a certain distance to close them. Even messenger-operation proved uncertain. Unless the sounding line hung nearly vertically the messenger might not slide down to the trigger until the bottle was well on its way back to the surface.

With the opening of the twentieth century, however, these difficulties were gradually eliminated, making use of greater precision engineering techniques, better design, new materials and the introduction of wire sounding lines for all apparatus.

REFERENCES AND NOTES

1. Howarth O 1931 *The British Association for the Advancement of Science: A Retrospect, 1831–1931* (London)
2. Forbes' influence on the British Association and its Dredging Committees has been explored by Rehbock P 1979 'The early dredgers: Naturalizing in British seas 1830–1850' *J. Hist. Biol.* **12**(2) pp293–368
3. Wyville Thomson C 1873 *The Depths of the Sea* (London)
4. Agassiz A 1888 *Three Cruises of the 'Blake'* (2 vols) (Washington)
5. Forrestier-Walker E 1952 *A History of the Wire-Rope Industry of Great Britain* (London)
6. Preece W 1861 'On the maintenance and durability of submarine cables in shallow water' *Min. Proc. Inst. Civil Engrs* **20** pp26–48
7. Wauchope R 1822 'Temperature of the ocean at great depths' *Mem. Wernerian Soc. Nat. Hist.* **4** pp164–5
8. Lenz E 1831 'Physikalische Beobachtungen angestellt auf einer Reise um die Welt unter dem Commando des Capitans Otto von Kotzebue in dem Jahren 1823–26' *Mem. Acad. Imp. Sci. St Petersburg* (*Sci-Math*) 6th ser, **1** pp221–341
9. Stevenson D 1872 *Canal and River Engineering* (Edinburgh) pp126–9
10. Denham H 1837 'On the tidal capacity of the Mersey estuary' *Rep. Br. Ass. Adv. Sci.* pp85–7
11. Wilke J C 1771 'Beskrifning pa et Instrument at Uphamta vatten ifran storre Djup utur Hafvet . . .' *Kong. Svensk. Vet. Akad. Hand.* pp60–6 & Plate 3
12. Biot J 1838 'Sur un moyen de puiser l'eau de mer à des grandes profondeurs . . .' *C.R. Acad. Sci. Paris* **7** pp303–6
13. Daubeny C 1838 'Instruments for taking a sample of water from considerable depths' *Trans. Soc. Arts* **51**(2) pp3–11
14. Rung G 1886 'Wasserschöpfer mit Tiefseethermometer' *Meteorol. Z.* **3** p549
15. Wyville Thomson C 1873 *Op. cit.*
16. Wyville Thomson C 1873 *Op. cit.* p300
17. Sigsbee C 1880 *Deep-sea Sounding and Dredging* (Washington)
18. Tanner Z 1896 'Deep-sea exploration: a general description of the steamer "Albatross", her appliances and methods' *Bull. US Fish Comm.* **16** pp342–3 and Plate 25

Chapter 8

Attempts to Perfect the Deep-sea Thermometer

8.1. THE BIMETALLIC THERMOMETERS OF SAXTON AND JOHNSON

The inherent weakness of the mercury-in-glass thermometer under compression led several inventors to consider the possibility of a bimetallic thermometer since a metal element was not itself susceptible to pressure. Such instruments contain a bimetallic strip composed of two metals having different coefficients of expansion so that any change of temperature causes the element to bend. This flexure drives a pointer which can be made to register its greatest degree of movement in either direction.

During the course of their Gulf Stream investigations begun in 1846, the US Coast Survey tried several types of thermometer and in 1860 Superintendent A D Bache presented a report on their merits.[1] Breguet's metallic thermometer, with a pressure-tight cover, gave good results and had been much used. An 'ordinary' thermometer, sealed into a glass globe, was also reliable but responded very slowly to temperature and could not be used at great depth. On these occasions a Six thermometer was sent down, although its register was often spoilt by rough handling.

Around 1848 Joseph Saxton, instrument-maker for the US Coast Survey, designed for them a thermometer based on that of Breguet. It was a large, elaborate and costly instrument (figure 96).

> Two stout ribands of silver and platinum are coiled and carefully united with silver solder, the more expansive metal in the interior. A coil of gold, to which each is soldered, serves to prevent the tendency of the silver and platinum to separate. The lower part of the coil is fastened to a brass stem, which passes up through the axis of the coil, while the opposite end to which the stem is attached is firmly fastened to the base of a short cylinder, which keeps that end of the coil from turning, and so gives the whole motion of the opposite end to the stem, which in turning indicates the turning of the coil from the relative expanding or contracting of the metals composing it. In this short cylinder are the wheels which serve to magnify the motion of the coil, a hand upon the dial indicating the motion, and pushing before it a registering hand, moving with sufficient friction to retain its place when thrust forward by the index hand of the thermometer. These instruments are graduated by trial. The brass and silver portions receive a thick coating of gold by the electrotype process, to prevent the action of seawater upon them.[2]

Bache found this thermometer 'admirable' and stated that 'it registered to ½° (say to 0.2° with certainty)'. However, careful experiments in a hydraulic press revealed that at pressures corresponding to 600 fathoms the error was indeed under 1 °F but increased to 7° or 9° at the equivalent of 1500 fathoms. Beyond that, it differed for each thermometer.

Maury himself thought that the two metals forming the element had differing compressibilities and would therefore be subject to an error which increased with depth. He concluded that the registered temperature would be too low, and this was in fact observed by Commander Rogers in the Arctic in 1855.[3] Rogers worked to only 40 fathoms but even at this depth his two Saxton thermometers consistently recorded lower

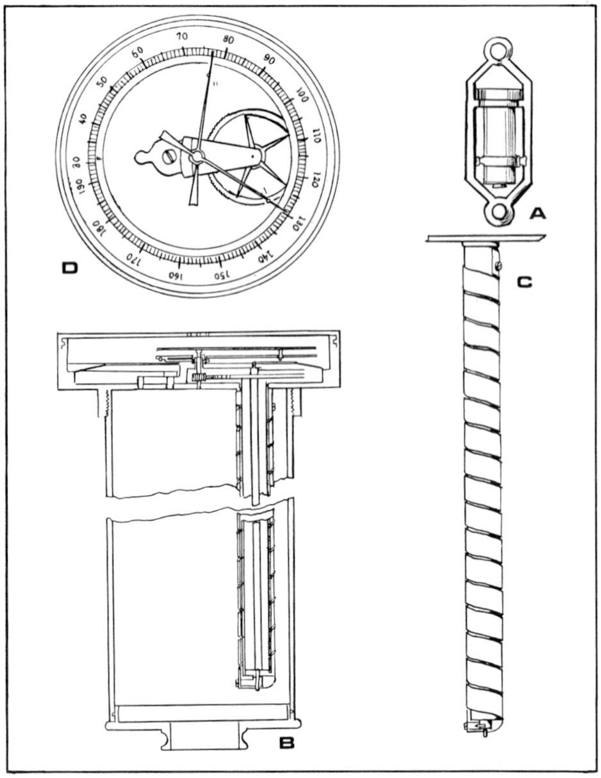

Figure 96. Saxton's bimetallic thermometer. A shows the instrument in its case, at a much reduced scale; B is a section through the thermometer; C, the bimetallic coil; D the dial. From Bache 1860 *Rep. US Coast Survey* Appendix 17, Fig. 4.

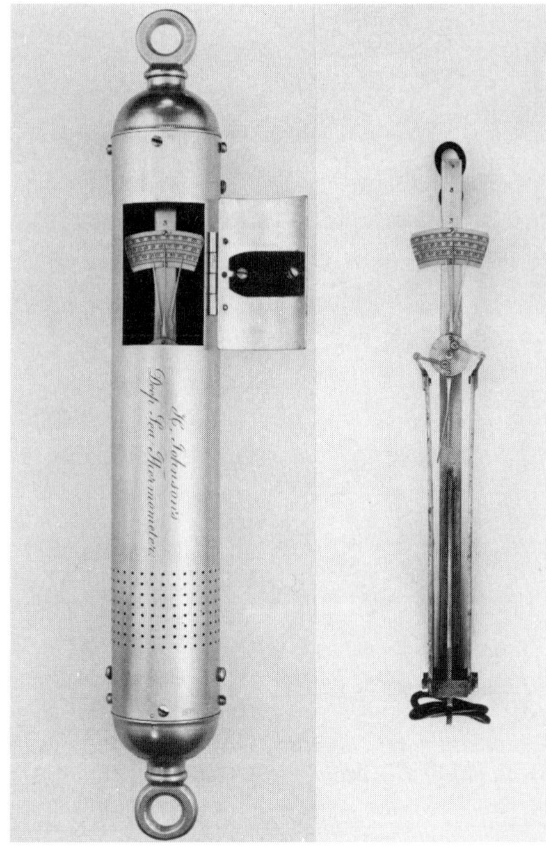

Figure 97. Johnson's bimetallic thermometer. *Left*, the instrument in its case; *right*, the frame, element and scale.

temperatures than he obtained with an insulated waterbottle; in fact, they differed from each other by up to 4 °F. Despite this, and their high cost ($70 each), Maury thought that they should be issued to ships on special survey duty.

The bimetallic thermometer designed by Henry Johnson in 1861 (figure 97) was claimed to be pressure-resistant by virtue of its flat elements.[4] These were:

> . . . thin bars of two metals rivetted together that expanded and contracted in different ratios with change of temperature. The . . . bars were composed of brass and steel, in the proportion of two-thirds of brass (which is the more dilatable metal) and one third of steel and had sufficient lateral motion to admit to legible scales of temperature, and also sufficient power to overcome the stiff friction of the indices.

The thermometer was carefully made, its pivot gilded and the bars tinned to resist corrosion. It was suspended on rubber bands within a pierced brass case, matching that of Johnson's pressure gauge for which it was intended to provide a temperature correction. The scale was fixed transverse to the elements so that the amount of travel, and hence the space allocated to each degree, was extremely limited. In use, the thermometer was found to be erratic, seldom in agreement with a mercury-in-glass thermometer, and generally unacceptable. In 1865 Captain J E Davis of the Admiralty Hydrographic Department tried it out again with Johnson's third pressure gauge.[5] It was offered for sale by Negretti and Zambra, and appeared in their catalogues of 1864, 1880 and 1885, but it did not prove popular.

Sir Joseph Prestwich, reviewing deep-sea temperature measurements made between 1749 and 1869, rejected all those from bimetallic thermometers.[6] Of the remainder he tried to discover which had been pressure-compensated, either by encasing the thermometer or by applying a correction to the observed reading. Certain anomalies stood out and these he took to be uncorrected in any way.

During the first half of the nineteenth century thermometers were shielded either by a sealed metal or glass container. One such metal case was illustrated by Aimé for the overflow thermometers used in his Mediterranean investigations (figure 98).[7] If such a case was air-filled it insulated the thermometer, whereas if it was filled beforehand with water there was no check that it had remained pressure-tight. From reports, it seems to have been difficult to make a sturdy case that did not leak or collapse, crushing the thermometer within. An unprotected instrument made of relatively elastic glass, on the other hand, deformed under pressure and returned to the surface unharmed.

In France, François Walferdin sheathed his overflow thermometers in evacuated glass tubes, a practice which he found more reliable than attempting to apply a pressure correction, which itself depended on a knowledge of the precise depth to which the thermometer had descended.[8]

Sir William Thomson reached a similar conclusion during his experiments to determine the effect of pressure on the freezing point of water,[9] but the effect was so minute that he had to take great pains to shield the thermometer lest pressure distort the reading beyond possible correction. Thomson used a sensitive ether thermometer about 10 inches long, sealed into an evacuated glass tube. Despite the insulation, when he put it into a tank of ice water and applied pressure, he noticed an immediate drop in temperature.

A straight-tube Phillips thermometer was used as a test standard by Casella during the development of the Miller–Casella thermometer in 1869. But despite familiarity with the sheathed thermometer, the complex shape of the Six instrument prevented its being treated in the same fashion. When eventually a straight-tube registering deep-sea thermometer was invented, towards the end of the nineteenth century, the glass sheath had become an integral part of its construction.

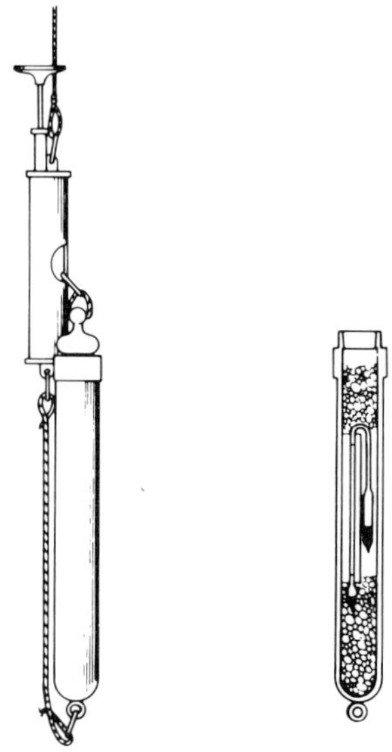

Figure 98. Aimé's thermometer. From Aimé 1845 *Exploration Scientifique de l'Algérie* vol 1, plate 4, Fig. 6.

8.3. SHIELDED-BULB THERMOMETERS: NEGRETTI AND ZAMBRA

Sheathing the entire thermometer reduced its sensitivity, but instrument-makers Negretti and Zambra claimed to avoid this with their pressure-protected Six thermometer first produced in 1857. It was developed as a result of the observations on the temperature of the River Thames made in 1844 by James Glaisher of the Royal Observatory, Greenwich. At a depth of 25 feet within the river additional pressure amounting to three-quarters of an atmosphere seriously distorted his thermometers. (This was the same series of observations that had led Henry Johnson to design the bimetallic thermometer referred to in §8.1.) Glaisher and Admiral Robert Fitzroy, then of the Board of Trade, approached Negretti and Zambra, and the evolution of the thermometer was described by them as follows:

> At the invitation of the late Admiral Fitzroy, various forms of thermometer were submitted to him, but it was found that no joints however well made, or cases no matter how strong, could resist the enormous pressure of the ocean at depths in which it was necessary to lower the instruments, and for a while it was almost thought that an instrument fulfilling the conditions required could not be constructed.
>
> Many were the experiments tried, till at length, in the year 1857, we made a thermometer whose bulb was hermetically sealed within an outer cylinder of glass, and the outer glass of such thickness that it was quite capable of resisting even greater pressure than was required: this thermometer we submitted to the late Admiral Fitzroy who at different times ordered about 50 for public service.[10]

Fitzroy explained how the manufacturers were able to combine protection with sensitivity:

> Negretti and Zambra undertook to make a case for the weak bulb, which should transmit temperature but reduce pressure. Accordingly a tube of thick glass is sealed outside of the delicate bulb, between which and the casing is a space all round, which is *nearly* filled with mercury. The *small* space not so filled is a vacuum into which the mercury can be expanded, or forced, by heat or mechanical compression, without doing injury or even compressing the inner or much more delicate bulb.
>
> This provision is meant to guard against possible compression of even the outer glass—strong as it is.
>
> One may ask, Why not strengthen the inner tube, the bulb, at once, so as to be equal in power of resistance to the outer casing? Mr Glaisher and the makers say, 'No, the bulb will yield a little on account of its length, be it even as strong as the outer case.'[11]

The 50 thermometers made to this pattern were supplied to several ships employed on sounding expeditions, among them HM Ships *Cyclops, Hydra, Medina, Firefly, Bulldog, Porcupine, Serpent, Gorgon, Rifleman, Swallow, Archer, Woodlark* and *Tartarus.* A thermometer issued to the steam yacht *Fox* was taken to the Arctic during the search for Franklin in 1859–60 and observations were made with it by Captain McClintock (figure 99).[12] Negretti and Zambra also applied the sheathed bulb to their deep-sea minimum thermometers, giving to these the non-corroding porcelain scales which they were fitting to other thermometers and barometers.

In use, both the Six and the minimum thermometers could be fitted into a slender, cylindrical valved case.

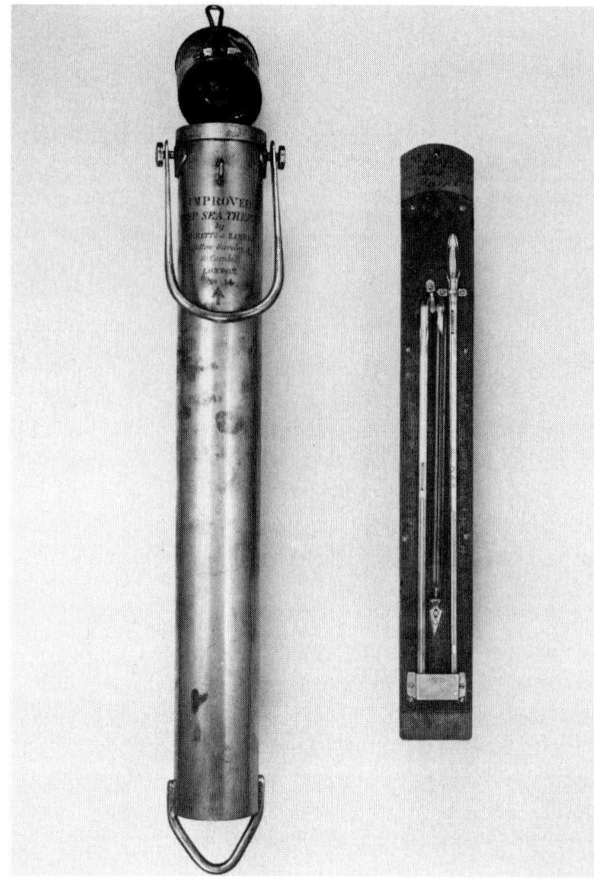

Figure 99. Negretti and Zambra pressure-protected thermometer (1857) with its valved Jamieson case (*left*).

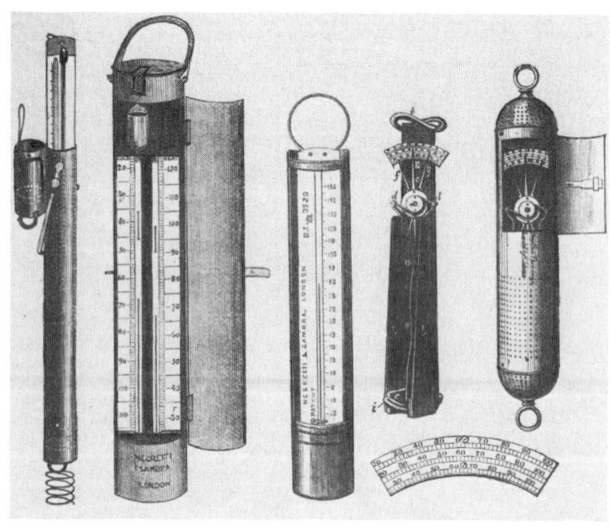

Figure 100. Deep-sea thermometers offered for sale in 1880 by Negretti and Zambra (*Illustrated Catalogue*, p147).

Those shown in Negretti and Zambra's catalogues (figure 100) differ from the surviving cases in which the Science Museum's thermometers are fitted (figure 99); these are of the pattern invented by Robert Jamieson of Glasgow in 1820.[13] Valves at the top and bottom of the case lifted during descent and closed when the apparatus was hauled in. The case was not strictly a waterbottle, although it did return full of water, since the cap was not tight-fitting and the sample may have been diluted on the way up.

8.4. SHIELDED-BULB THERMOMETERS: MILLER–CASELLA

The standard Hydrographic Office pattern Six thermometers were much shorter than the Negretti and Zambra version; some six inches overall, three-limbed but mounted on an ebonite board with only the outer limbs visible. Ebonite, a mixture of two-thirds rubber and one-third sulphur heated to 150 °C, had been invented in America in 1833 and was rediscovered independently in Britain about ten years later. The thermometer was carried in a copper cylinder pierced at the top and bottom, with a little hinged door opposite the scales so that it could be read without taking it from the case. During the *Lightning* cruise Wyville Thomson found that his instruments were 'very wild at a few hundred fathoms'.[14] Consequently, when the cruise was ended, the deep-sea committee of the Royal Society met at the Hydrographic Office where the Society's Vice-President, Dr W A Miller, set out proposals for a pressure-resistant thermometer which he subsequently described to the Society in June 1869.[15]

Made by the firm of Casella (figure 101), its outward appearance resembled the Hydrographic Office pattern, but it now had a simple U-shape, the bulb was reduced in size and was continuous with the tube. A smaller bulb on the opposite limb held the vapour whose elasticity returned the fluids as the temperature fell. The bulb was surrounded by a glass sheath, while the intervening space was filled with alcohol, its vapour and some air. Since the level of mercury and the indexes were not clearly seen against the dark ebonite mount, the tube was backed with white glass and the scales engraved on slips of white glass fastened beside the tube, in similar fashion to the HO pattern thermometer. Small bulges or aneurisms at the ends of each limb prevented the indexes being carried into the bulbs or down into the U-bend.

There was certainly a need for an improved instrument; the cumbersome Fitzroy–Negretti thermometer was rather fragile and since it never overcame the problem of shifting indexes its production was soon discontinued.[16] Curiously, Miller denied all knowledge of this earlier thermometer, despite the number of surveying crews who must have used it, and when his 'new invention' was announced, Negretti and Zambra made strenuous objections to his claim. Figure 102 compares the two designs. They themselves had failed to patent their improvement and they may have recognised that it was not new, having been preceded by other sheathed thermometers,

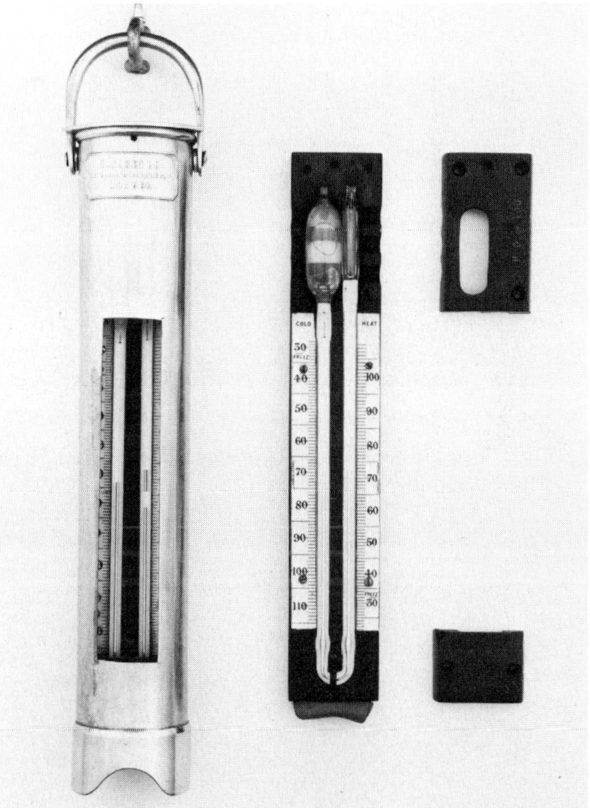

Figure 101. Miller–Casella pressure-protected thermometer, 1869. *Left,* assembled in its copper case; *right,* dismantled to show construction.

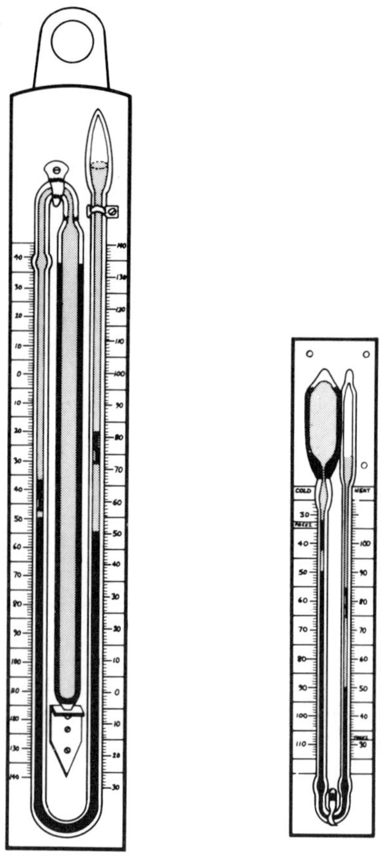

Figure 102. Comparative size and design of the Negretti and Zambra and Miller–Casella pressure-protected thermometers.

though perhaps not intended specifically for marine use. However, Negretti wrote to the editor of *Nature* in 1873,[17] drawing attention to the description printed in his *Treatise* of 1864,[18] and another letter was written by R H Scott FRS on their behalf, reminding readers that Admiral Fitzroy had exhibited a Negretti thermometer to the Royal Services Institution in 1859.[19] Resentment was still felt many years later, and Negretti's 1880 catalogue restated their case:

> One thing we will assert, and that is, that neither Dr Miller nor his assistant inventor of the now celebrated thermometer would have dared exhibit the thermometer as a novelty, had Admiral Fitzroy been alive.[20]

This 1880 catalogue illustrated the range of thermometers that Negretti were offering at this date (see figure 100): Johnson's bimetallic, old forms of the Six and deep-sea minimum thermometers, Negretti's own version of the Miller–Casella and the new reversing thermometer which had by then superseded these antiquated instruments for most purposes.

Whatever its defects, the Miller–Casella thermometer represented Six's concept carried as near to perfection as was possible. Casella made strenuous efforts to measure the effect of pressure on his instruments, building a compression-chamber for test purposes and comparing his own with thermometers supplied by Pastorelli and Elliott (Negretti and Zambra declined to send him any of their own manufacture). Such tests did not truly parallel deep-sea conditions, however, and under compression the materials composing the thermometer heated up considerably. In the sea this heat was dissipated as fast as it built up, but within the confines of a test-chamber it required a long period of time and careful comparison with a standard thermometer to ensure that this heat was lost.[21]

Despite his care, however, Casella found that no correction table could be drawn up which held good for all thermometers (as indeed Walferdin had found some 30 years previously). Though possible theoretically, in practice, individual characteristics of manufacture meant that a table had to be compiled for each thermometer, and this in turn required that each be lowered to a known depth for testing and comparison, a requirement that was still difficult to fulfil, given the technical problems of sounding.

8.5. A NEW APPROACH: THE REVERSING THERMOMETER

Negretti and Zambra patented a new reversing thermometer in January 1874 (figure 103) in which they abandoned the defective Six instrument, and substituted a new form which borrowed features already proven in medical and industrial registering thermometers. Whereas the Six thermometer was a complex, wide-bore tube containing two liquids and two indexes, the new reversing thermometer contained only mercury and had no indexes. Some trials were made with a U-shaped tube but soon the straight-tube

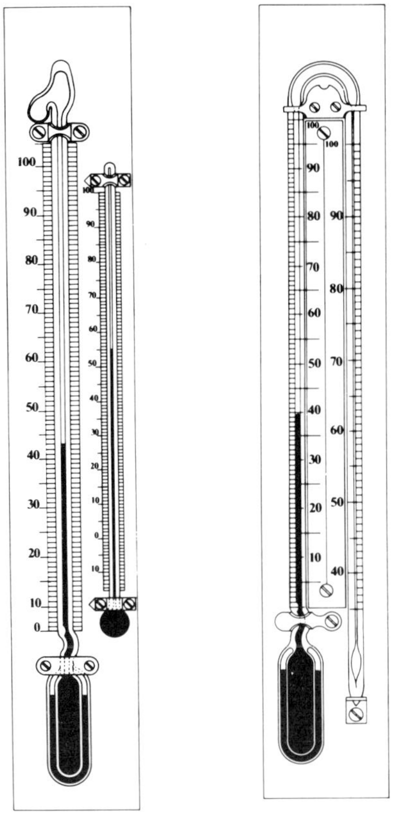

Figure 103. Patent drawings of Negretti and Zambra reversing thermometers, straight and U-forms, 1874 (*GB Patent Specification* No. 109).

form, within a glass envelope, proved superior in every respect.

This improved instrument thus avoided the inherent weaknesses of the Six thermometer—the tendency of the alcohol to 'creep' past the mercury segment, and the looseness of the indexes which came about when the alcohol destroyed the springiness of their bristles. The Six thermometer had its scales placed beside the tube, but on the new model, protected by an outer envelope, fine divisions could be engraved directly on the stem, thus diminishing errors of parallax made when reading the instrument. Finally, whereas the Six thermometer registered the extremes of temperature through which it passed, irrespective of their depth, the reversing thermometer registered a single temperature, that at which it was reversed, and was unaffected by any subsequent heating or cooling.

Unlike the Six thermometer which demanded careful handling, Negretti's thermometer was to be abruptly capsized at the depth being measured, as the patent specification explained:

> . . . we form an obstruction in the thermometer (by preference at or about zero near the bulb) which we term the 'breaking point' and at the top of the bulb we place a small reservoir which may be in the form of a tube or a prolongation of the thermometer tube, or it may consist of a bulb into which a portion of the mercury may be run, and for the purposes hereafter explained. The instrument is mounted on a centre pin or in a case or frame in which it may be turned so that its position may be reversed when required.

It was made to register

> . . . by turning it over on its centre pin . . . the amount of mercury at the time above the 'breaking point' will break away . . . and will flow to the top of the tube whence it will run into the small reservoir . . . where it will remain when the tube resumes its normal position.[22]

A second thermometer on the backplate was read before and after lowering, the difference between the two readings representing the volume of mercury that flowed into the upper chamber at the moment of reversal, and thus indicating the temperature at that moment. It was assumed that the temperature fell below the surface, otherwise the thermometers could be cooled or warmed before they were lowered.

The patent also showed a U-tube thermometer. To register, it had to be rotated through 360° whereupon the mercury above the constriction flowed round the tube and came to rest in the opposite limb, to be read off directly against an appropriate scale. Some of these thermometers, together with frames, were sent for trial aboard HMS *Challenger* (figure 104). As the frame sank its vane turned freely, but on being drawn upwards through the water it engaged a pinion which caused the thermometer to overturn and lock after one complete revolution. It was found in practice that this lock brought such a strain on the cog that it often sheared off its spindle, so that Tizard, the staff officer responsible for the trials, had the defect remedied by Ferguson, *Challenger*'s chief engineer, who

> . . . applied an ingenious apparatus by which when the thermometer has made one complete revolution, the pinion is lifted clear of the cogwheel, and thus the propeller is allowed to revolve as freely in its ascent as it did in its descent.[23]

Captain Nares took the same 'ingenious apparatus' when in 1875 he was detached from command of *Challenger* to lead the Arctic expedition in HMS *Alert*

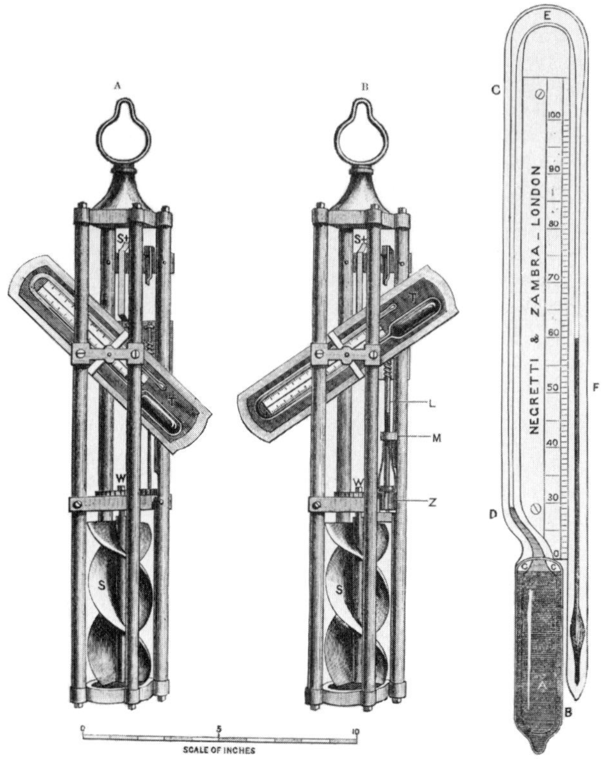

Figure 104. Negretti and Zambra reversing thermometer and frame, as used on *Challenger.* The centre figure shows the modification added by Ferguson (marked LMZ). From *Challenger Report Narrative* 1885 vol. 1, Pt 1, Fig. 29.

(figure 105 shows one of his thermometers).

It was more straightforward to make the thermometer turn through 180°, assisted by a mercury-filled sheath on the bulb which made the instrument unstable about its centre. There was an advantage in the straight-tube design in that it could be contained in a glass envelope, so protecting the tube or upper bulb from the effects of pressure. This feature was not patented but it was fully described in *Engineering* magazine in 1878 (see figure 103 *left*):

> By the simple expedient of placing the thermometer entirely in a shield of glass hermetically sealed the effect of external pressure is entirely eliminated. The shield must of course be strong. It need not be exhausted of air. It must however render the enclosed thermometer more difficult to be affected by changes of temperature; in other words it will make it sluggish. To counteract this sluggishness, in that portion of the shield surrounding the bulb, some mercury is introduced and confined there by a partition cemented in the shield around the neck of the thermometer bulb.[24]

Kew Observatory tests showed these instruments to be quite free of distortion. They could be read with great precision because within the thin-bore tube the mercury travelled further for each degree of temperature change and the graduations were cut directly on the stem and thrown into relief against white glass backing the tube (figure 107 *centre*).

As a simple means of overturning a straight thermometer, Negretti proposed a hollow wooden block weighted by loose lead shot (figure 106), but this had to be hauled up smoothly and was only practical in calm, shallow water (figure 107). A compact small reversing frame was marketed by the company in 1885,

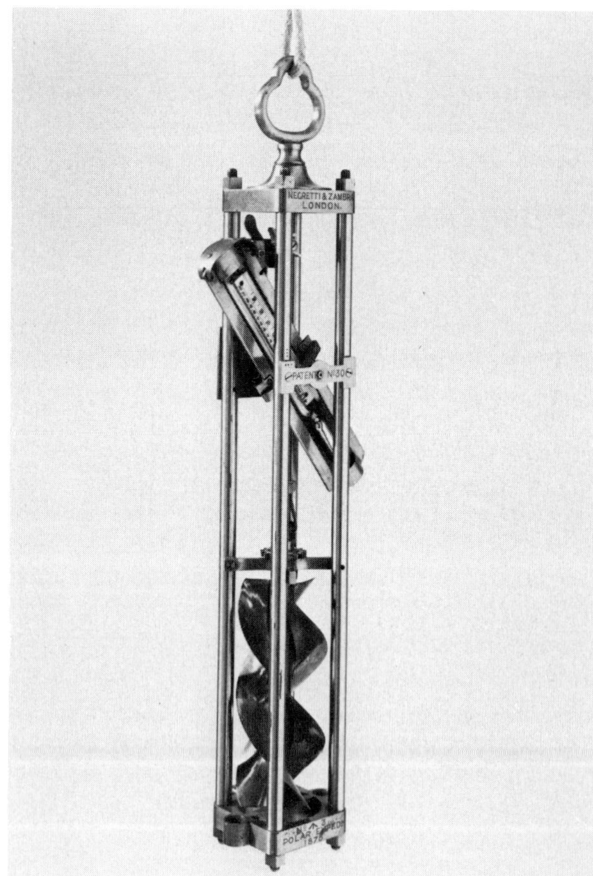

Figure 105. Challenger's reversing thermometer, also used on the 1875 Arctic expedition.

credited to Admiral Magnaghi, Hydrographer to the Royal Italian Navy (figure 108), which was probably devised during the cruise of the Italian ship *Washington.*[25] Inspired perhaps by Sigsbee's popular waterbottle, a propeller in an upper part of the frame turned as the line was brought in, releasing the thermometer which then capsized about its axis. About 10 to 15 fathoms of lift were needed to turn the thermometer—an excessive distance in water where the temperature changed rapidly with depth. Testing the frame on board *Triton* in 1882, and on Buchanan's ship *Mallard,* a rubber band was wound round the frame to make the reversal faster, but in really shallow water instant overturning was essential and this could best be done by messenger operation.[26]

H R Mill of the Scottish Marine Station at Granton also designed a suitable frame of this type in 1884 (figure 109):

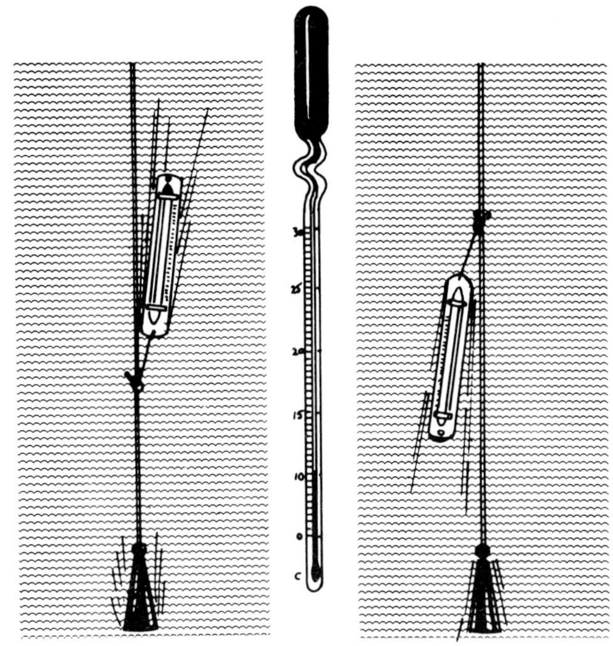

Figure 107. Negretti and Zambra reversing thermometer shown with a simple means of reversing it on a wooden block as the line is hauled up.

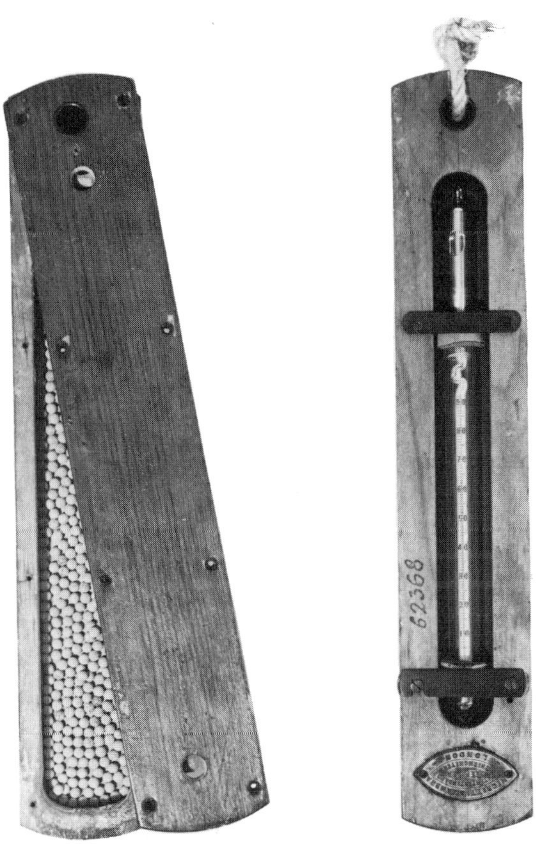

Figure 106. Block for overturning the Negretti and Zambra thermometer in shallow water.

> The Scottish thermometer frame, as the improved form has been named, differs from Magnaghi's by the removal of the screw and fan from the pin that holds the thermometer in position, and the substitution in their place of a lever kept down by a spiral spring. The lever is forked at the outer end to allow the line to which the thermometer is attached to pass between. In order to prevent the line from slipping out of the fork, it is clasped by two thin springs which rise from a projecting horseshoe-shaped piece beneath the fork. A hollow cylindrical weight slightly belled-out at the lower end is dropped down the line when the thermometer is to be reversed; it falls on the forked lever, raises the pin, the upper end of the instrument gets an outward impulse from an india rubber band slipped over the frame (a device of Mr Buchanan's) and it turns sharply, and is clamped by a brass spring.[27]

This action could in turn release another messenger attached to the frame, to trigger the frame below, and so on down the line. The Scottish pattern frame was not new, Mill added, but improved on a wooden frame made by Captain Rung of the Danish Meteorological Service, which, in Mill's view, was not wholly satisfactory.

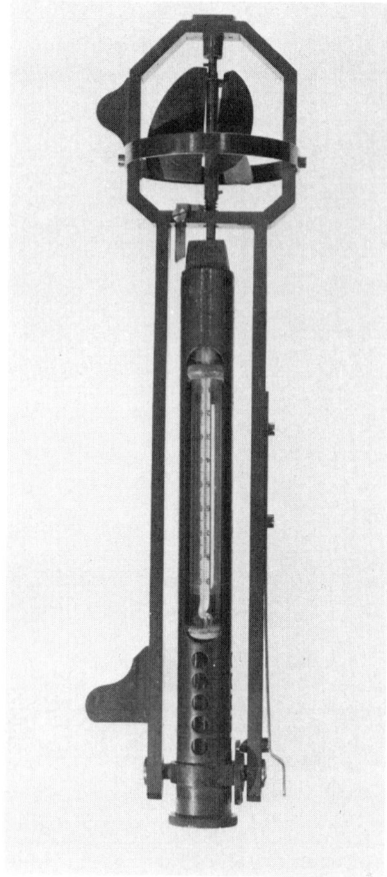

Figure 108. Negretti and Zambra thermometer fitted in a Magnaghi reversing frame.

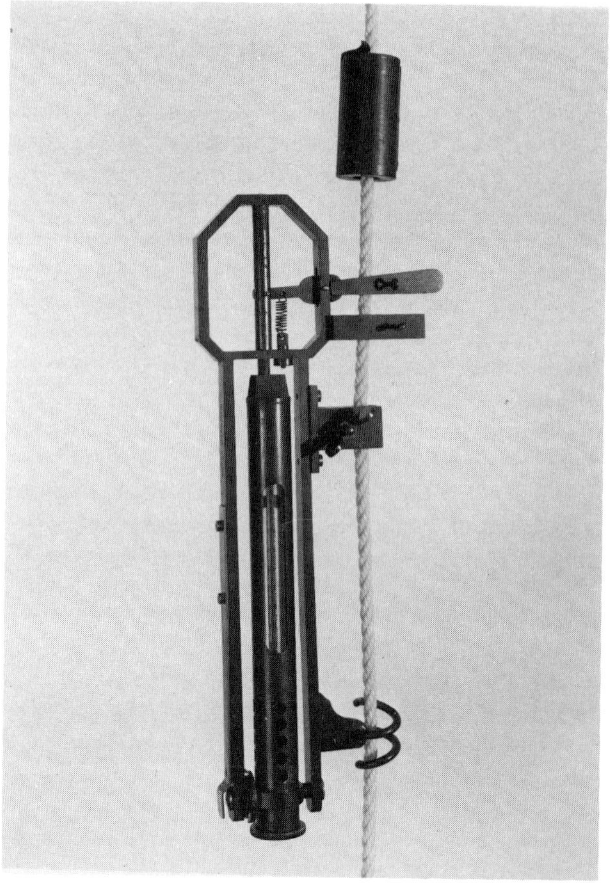

Figure 109. Negretti and Zambra thermometer fitted in a 'Scottish pattern' reversing frame.

8.6. SIEMENS' ELECTRICAL RESISTANCE THERMOMETER

The electrical resistivity of a conductor is governed by its temperature, and, given this physical law, it is possible to measure changing resistance and thus changing temperature. In 1866 Werner and William Siemens described such a thermometer, developed from an earlier industrial apparatus which they had adapted to work in the deep sea, consisting of two resistance coils, one lowered into the sea and the other placed in a control water tank on deck. The circuit was taken across a Wheatstone bridge and balanced by adjusting the temperature of water in the control tank until it equalled that around the lowered coil. A galvanometer indicated when the bridge was so balanced and a test thermometer in the control tank registered the temperature.[28]

Later, the Siemens brothers developed a generator to replace the original batteries, as well as an elaborate method of adding and mixing refrigerant to the control tank. With care it was possible to lower the coil and to take simultaneous readings, thus giving a detailed temperature profile. In view of later difficulties, however, it seems probable that the tests on which this claim was based were carried out in harbour.

The Siemens differential thermometer was tried out on *Porcupine* during her 1869 cruise,[29] and 1000 fathoms of small cable, 'about the size of ordinary sounding line', but containing two insulated wires, were taken on board. The resistance coils were of fine platinum wire and a freezing mixture of ice was intended to cool the control tank. At sea, however, the ship's motion kept the galvanometer in constant movement so that it could not be read, and so the apparatus was therefore not able to fulfil its intended purpose, which was to resolve the level at which the coldest water lay.

In 1870 Siemens provided an improved galvanometer but, on the declaration of war between Germany and France, the German mechanic trained to operate the equipment was recalled home and in these circumstances it was considered best to leave the apparatus ashore.[30] The differential thermometer was however issued to the *Challenger,* a Thomson marine galvanometer being taken to show the balance of the bridge. Several more or less successful observations were made which corresponded with those from protected thermometers. Since no permanent site had been fitted up for the apparatus there were no regular observations.

Some years later Commander Bartlett of the US Coast and Geodetic Survey (as it had then become) was ordered to test the thermometer, and he installed it on the *Blake* (figure 110).[31] There were some difficulties after the preliminary set-up and tests because the Carré ice-machine ordered from France had failed to arrive. This hand-operated machine was essential for providing chilled water for the control tank. Eventually a 'common ice-cream freezer' was procured, to make ice ashore, and *Blake* moved into the Gulf Stream, running a profile over previous soundings. A series of observations were taken using Miller–Casella thermometers immediately prior to the Siemens' measurements. Five minutes were allowed for the coil to assume the ambient temperature, and after balancing the bridge, another five minutes were allowed to elapse to ensure the stability of both coils. In the majority of cases the results were identical to those made by the Miller–Casella thermometers.

Bartlett also found that the vessel's roll affected the mirror galvanometer but since the deflection was symmetrical it was easy to determine the middle point. The mirror was seriously disturbed by the vibration of the ship's engines, and so these had to be stopped when the apparatus was in use. This made it very difficult to maintain station and keep a vertical sounding line within the strong Gulf Stream current.

Some days later, after the ice-machine had arrived, another trial was made. The first series of tests had to be made in daylight since the cable could not be seen after dark, but Bartlett must have had a light source brought aboard, and was then able to work through the night for the second series, taking temperatures at the surface and at 17 levels down to 400 fathoms and back, each reading taking about 15 minutes. The following morning the series was repeated. Correspondence with the Miller–Casella readings was excellent and both instruments showed the anomalous warm layer at 50 fathoms. In order to register this 'high' on the Miller–Casella thermometers they were lowered rapidly to that depth and after eight minutes reeled in at 200 fathoms per minute so that the minimum side did not have a chance to register a lower temperature. Such an operation supposed a previous knowledge of the existence of such a layer.

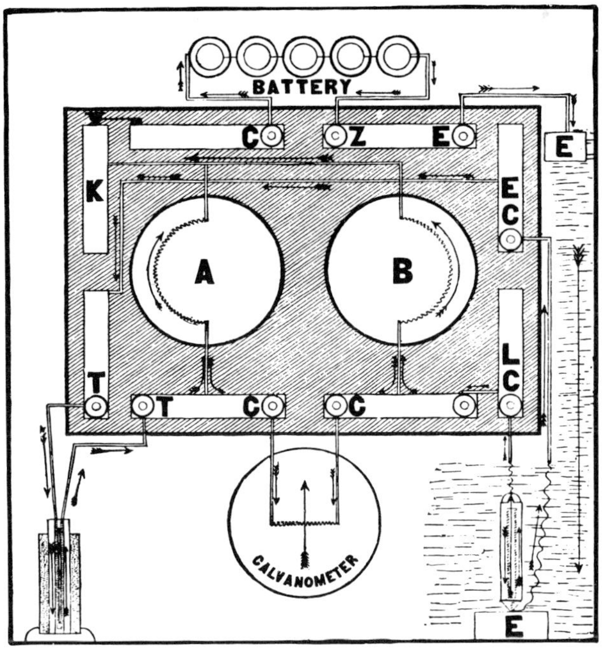

Figure 110. Siemens' electrical thermometer as arranged on board USS *Blake*. From Agassiz 1888 *Three Cruises of the 'Blake'* vol. 1, Fig. 15.

Bartlett's report was favourable as regards the thermometer itself, but not on the means of working it.

> The cable containing the insulated wire is very loosely laid up, and in the few lowerings that we have made it has become long-jawed†. The parcelling round the cable is very poor and the slightest chafe rubs it off. The apparatus could not be used constantly as the constant paying out and reeling in would soon part the copper wires, but as an instrument of precision I consider it a great success. More

† Long-jawed = tending to unlay.

confidence may now be placed in the Miller–Casella thermometers, and at certain times on each line the electrical apparatus could be used to verify the temperatures and show any underlying strata of water. All parts of the apparatus worked to perfection but it is necessarily very slow work taking a serial to any great depth.[32]

The *Challenger Report* comments, written after Bartlett's trials, were also favourable although her scientists had made little use of the apparatus:

> When accurate temperature observations are required from intermediate depths, this instrument is especially valuable and it will in all probability be extensively used in deep-sea observations.[33]

This forecast was not borne out, however. The Siemens brothers brought Bartlett's report to the notice of the Royal Society without mentioning (or perhaps they were unaware of) the problems of repeatedly working a deep-sea line containing an electrical conductor.[34] In fact, by this time such observations could be more speedily and conveniently carried out with several reversing thermometers spaced along a single sounding-line, thus avoiding the need for ice-machines, galvanometers or the ship's being hove-to with engines stopped for several hours.

At the end of the nineteenth century *Valdivia,* cruising in the southern hemisphere in 1898–99, tested a 'telethermometer' whose resistance coils were measured with a Wheatstone bridge and a telephone. But in deep water a galvanometer still had to be used and this was never satisfactory on a moving ship. Additionally, on a long cable submerged to great depth, there was a tendency for the insulant to fail. In order to facilitate their reeling in and out, such cables were slender and the insulating material thin and fragile.[35]

To summarise the various methods that were tried during the nineteenth century to measure deep-sea temperatures, the first, collecting water in an insulated container, proved far from ideal for it was attempted before suitable insulating materials were available. About 50 years later the technology was to hand and then it was tried again, with great success. Bimetallic thermometers were also a failure; their working parts were defective and they suffered from unquantifiable pressure errors. The Six thermometer was somewhat improved but it was not capable of the precision demanded and it always suffered from loose indexes. A new liquid-in-glass thermometer could be made to high standards of accuracy and has survived to the present day. The use of electricity to telemeter the register from probe to ship proved cumbersome and slow to handle, and so was abandoned by the turn of the century.

REFERENCES AND NOTES

1. Bache A D 1860 'Lecture on the Gulf Stream prepared at the request of the American Association for the Advancement of Science' *Rep. US Coast Survey* Appendix 17 pp165–8
2. Bache A D 1860 *Op. cit.* p166
3. Maury M F 1858 *Explanation and Sailing Directions to Accompany the Wind and Current Charts* 8th edn (Washington) pp125–6
4. Glaisher J 1861 'On a deep-sea thermometer invented by Henry Johnson Esq.' *Rep. Br. Ass. Adv. Sci.* Sections pp58–9
5. Glaisher J 1866 'Experiment off Ventnor with Mr Johnson's deep-sea pressure gauge' *Rep. Br. Ass. Adv. Sci.* Sections pp24–7
6. Prestwich J 1875 'Tables of temperatures of the sea at different depths . . .' *Phil. Trans.* **165** pp587–672
7. Aimé G 1845 *Ann. Chim. Phys.* 3me ser. **15** Plate 1, fig. 6
8. Walferdin F 1839–40 'Mémoire sur les effets de pression et sur d'autres causes d'erreur qui peuvent affecter les observations de temperature faites sous l'eau et à des grands profondeurs' *Bull. Soc. Geol. France* **5** pp83–93
9. Thomson W 1844–50 'The effect of pressure in lowering the freezing point of water experimentally demonstrated' *Proc. R. Soc. Edinburgh* **2** pp267–71
10. Negretti and Zambra 1880 *Encyclopedic Illustrated and Descriptive Catalogue* pp509 (London).
11. Great Britain Board of Trade 1857 *Meteorological Papers No.* 1, p55
12. McClintock L 1859–60 'Report of scientific researches made during the late voyage of the yacht "Fox" in search of the Franklin expedition' *Proc. R. Soc.* **10** pp148–51
13. Jamieson R 1821 'Description of the marine thermometer case invented by Robert Jamieson of Glasgow' *Phil. Mag.* **57** pp294–300
14. Wyville Thomson C 1873 *The Depths of the Sea* (London) p290
15. Miller W A 1869 'Note upon a self-registering thermometer adapted to deep-sea soundings' *Proc. R. Soc.* **17** pp482–6

16. Negretti and Zambra 1885 *Encyclopedic Illustrated and Descriptive Catalogue* (revised and corrected edn) (London) p173
17. Letter dated 14 October published in *Nature* **8,** 23 October 1873, p529
18. Negretti and Zambra 1864 *A Treatise on Meteorological Instruments* (London) p90
19. Letter dated 9 December published in *Nature* **9,** 11 December 1873, p102
20. Negretti and Zambra 1880 *Op. cit.* p510
21. Captain J E Davis submitted a report giving full details of these tests in Davis J E 1871 'On deep-sea thermometers' *Proc. Meteorol. Soc.* **5** (55) pp305–42
22. *GB Patent Specification* No. 109 (1874)
23. *Challenger Report Narrative* 1885 **1** (1) p89
24. Anon 1878 *Engineering* (March issue) p226
25. Buchanan J Y 1883 'Mediterranean', in *Encyclopaedia Britannica* p823; and Campigli P 1902 'Note biografiche sul vice-ammiraglio Magnaghi' *Rev. Geog. Italiana* **9** (10) p610
26. *Challenger Report* 1885 *Op. cit.* p95
27. Mill H R 1883–4 'On the periodic variation of temperature in tidal basins' *Proc. R. Soc. Edinburgh* **12** pp928–9
28. *GB Patent Specification* No. 2181 (1861). This covered an industrial differential thermometer, the deep-sea adaptation was given in Siemens C W & W 1886 'Methode für fortlaufende Beobachtungen der Meerestemperatur bei Tiefen messungen' *Poggend. Ann. Phys. Chem.* 5th ser. **9** pp647–9; and Siemens C W 1869 'Thermomètre éléctrique balance pour measurer la température du fond de la mer' *Les Mondes* **20** pp269–72
29. Carpenter W, Jeffreys J G and Wyville Thomson C 1869 'On the scientific exploration of the deep sea' *Proc. R. Soc.* **18** pp411–12
30. Carpenter W and Jeffreys J G 1870–71 'Deep-sea researches' *Proc. R. Soc.* **19** p151
31. Bartlett J 1882 'Report on the Siemens electrical deep-sea thermometer' *Rep. US Coast and Geodetic Survey* Appendix 18 pp451–7
32. Bartlett J 1882 *Op. cit.* p455
33. *Challenger Report* 1885 *Op. cit.* p97
34. Siemens C W 1882–83 'A deep-sea electrical thermometer' *Proc. R. Soc.* **34** pp89–95
35. Gasnier P 1901 'Thermomètres de profondeur' *La Géographie* **4** pp168–71

Chapter 9

The Cruise of HMS Challenger, *1872–76*

9.1. PLANNING THE VOYAGE

By 1870 a true science of the sea had come into being. In 1865 Professor Forchammer of Copenhagen published the results of 20 years' patient work analysing sea water from all parts of the world. American and British vessels ran sounding surveys around the continental margins and over those ocean crossings of commercial interest, but not, as yet, the transpacific crossing to Australia. From 1867 American biologists went each summer to the Gulf Stream while the British ships *Lightning* and *Porcupine* showed what a modest programme and reliable apparatus could achieve. A new concept of the oceans took shape, in which the whole body of water gradually circulated and throughout which some forms of life would be found despite extreme conditions.

In 1871 Dr Carpenter drew the Royal Society's attention to the large-scale oceanographic investigations that other nations were now formulating, and urged that Britain should participate in this work. The Austro–Hungarian government had already organised a cruise by the *Novara* between 1857 and 1859. The *Novara* was sent to 'show the flag', but also undertook a modest scientific programme. In 1862 the results of the voyage were published in 24 volumes, thereby making available to the scientific community all the data so carefully gathered.

The Royal Society proposed that a ship and crew, supplemented by a scientific staff, should visit several oceans and extend knowledge of the deep sea. A comprehensive programme was drawn up, listing the desirable observations and the theories which they were intended to test. Physical observations were to cover sounding and bottom sampling, surface and submarine temperatures, tides and currents, meteorology and terrestrial magnetism. It was also hoped to collect evidence of a slow, deep-water drift from the poles to the equator; such an abyssal circulation would balance the observed global surface current system.

The chemists were to analyse sediments and sea water, and the gases collected from the water and from swim-bladders of fish. The greatest labours fell to the botanists who were to gather all plant material, from the marine algae of the open sea, through seaweeds to the vast number of land species encountered on expeditions ashore. It was stressed that it was likely that many remote islands, as well as certain unexplored coasts of the great continents, possessed flora unknown to science.

The scientific director of the expedition, Professor Wyville Thomson, was an accomplished zoologist, and the instructions under this heading allowed him more discretion. Two points were made, namely that *Challenger*'s route should take her through the little-known northern and eastern Pacific, and that hydrographic observations should be made on both sides of Wallace's Line, that important faunal divide within the Malay Archipelago. Seeking to leave no stone unturned, literally and figuratively, the recommendations reminded the shore parties to investigate the geology and anthropology of these regions.

The voyage of *Challenger* was the culmination of the research backed by the Admiralty and the Royal Society. The ship visited every ocean except the Arctic and probed into every aspect of the sea with the exception of deep currents which could not yet be measured. Whenever possible, the naturalists went ashore to extend their observations on land and in the rivers, and to meet the native peoples. Surveys were made, and magnetic and meteorological registers kept throughout the voyage. However, because of the anticipated long absence from her home base it was decided that *Challenger* should not rely on novel apparatus whose failure might jeopardise the expedition's chance of success. The Admiralty therefore, while it issued 'all the instruments and apparatus which modern science and practice have been able to suggest and devise', also included recent inventions in addition to, rather than as replacements for, the older familiar gear (figure 111).[1] *Challenger* was voyaging into literally uncharted waters. Nonetheless the colonies were not wanting in skilled instrument-makers; the Navy seems to have been conservative by nature. From the descriptions of apparatus and laboratories which survive it is clear that many instruments were altered and constructed on board ship or in port, and that the layout and methods of work in the laboratories were also changed in response to climate as the ship moved from the tropics to polar regions, as well as simply to improve working standards (figure 112).

NARRATIVE OF THE CRUISE. 41

APPENDIX C.

STORES AND APPARATUS SUPPLIED FOR THE USE OF THE EXPEDITION.

APPARATUS FOR DEEP-SEA SOUNDINGS, TAKING TEMPERATURES, AND PROCURING SPECIMENS OF THE WATER.

Name of Article.	Quantity originally supplied.	Additional quantity received in the course of the voyage.	Total quantity expended.	Remarks.
Sounding rods, Hydra,	18	...	10	
,, Baillie,	...	12	2	
Sinkers, 1 cwt.,	240	...	228	
,, $\frac{1}{2}$ cwt.,	...	1,479	1,281	
Valve leads, $1\frac{1}{2}$ and 1 cwt.,	...	30	2	
Cup leads, 1 cwt.,	40	...	4	
Sounding line, No. 1, (fathoms)	20,000	20,000	26,000	
,, No. 2, ,,	64,000	...	34,000	
Iron gin-blocks, patent sheaves, 9 inch,	3	...	...	
Worsted for marking lines, (lbs.)	50	60	70	
Accumulators,	50	...	50	
For suspending sinkers, Wire, (fathoms)	2,240	...	1,000	
,, ,, Washers,	730	...	430	
For deep-sea sounding. Wire, (fathoms)	10,000	...	...	
For deep-sea sounding. Drums,	2	...	...	
For deep-sea sounding. Clock-work,	2	...	...	
Thermometers (protected),	35	69	48	
Slip water-bottles,	6	12	12	
Stop-cock water-bottles,	6	6	6	
APPARATUS FOR TRYING SUBMARINE CURRENTS.				
Current line, (fathoms)	60,000	...	23,000	
Current buoys,	25	...	7	
Current drags,	2	...	2	
APPARATUS FOR DREDGING AND TRAWLING.				
Dredges,	34	...	11	
Dredge-nets,	20	...	11	
Trawls,	22	...	16	
Rope, 3 inch, (fathoms)	10,000	...	4,200	
,, $2\frac{1}{2}$ inch, ,,	10,000	21,000	27,100	
,, 2 inch, ,,	5,000	10,000	10,860	
Accumulators,	100	160	260	
Iron gin-blocks, patent sheaves, 12 inch,	6	...	...	
,, ,, 9 inch,	3	...	...	
Copper sieves, (sets)	6	...	3	
Copper wire ladles,	7	...	4	
Swivels,	100	...	40	
Galvanized iron chain, $\frac{1}{4}$ inch, (fathoms)	280	...	120	

ORDINARY SOUNDING GEAR SUPPLIED.

Article.	Quantity.	Article.	Quantity.	Article.	Quantity.
Deep-sea leads,	50	Twine,	600 lbs.	Tents,	5
Hand leads,	100	Beeswax,	24 lbs.	Boats' cooking stoves,	5
Cup leads, $\frac{1}{2}$ cwt.,	30	Boats' compasses,	6	American axes,	12
Patent logs,	24	Boats' sounding davits,	9	Hand saws,	12
Burts' nippers, and bags,	12				

Figure 111. Stores and apparatus taken on board *Challenger*. From *Challenger Report Narrative* 1885 vol. 1 Pt 1, pp 41–2.

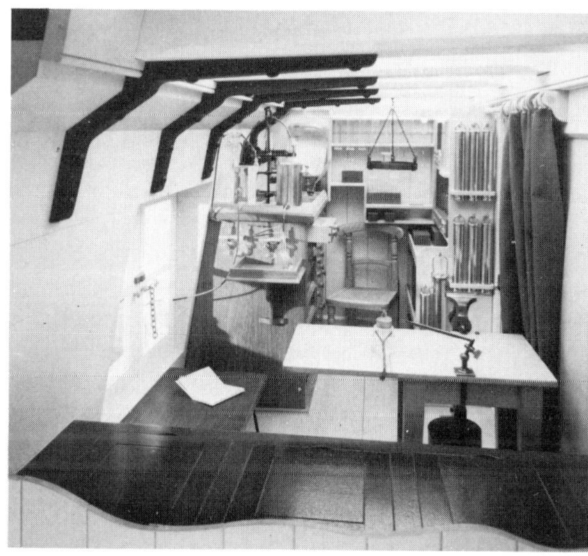

Figure 112. (*right*) Model of *Challenger*'s chemistry laboratory.

9.2. THE APPARATUS AS SHIPPED

The *Challenger*'s soundings were made from platforms projecting from each side of the ship abreast the foremast. For this operation the sails were furled and the ship brought head-to-wind under steam. The sounding line was carried through an accumulator and various sounders were attached, according to the anticipated depth to be plumbed (figure 113). When this was less than 1000 fathoms, a tube lead or cup lead

Figure 113. Model of the bow section of HMS *Challenger* showing the sounding gear rigged ready for use.

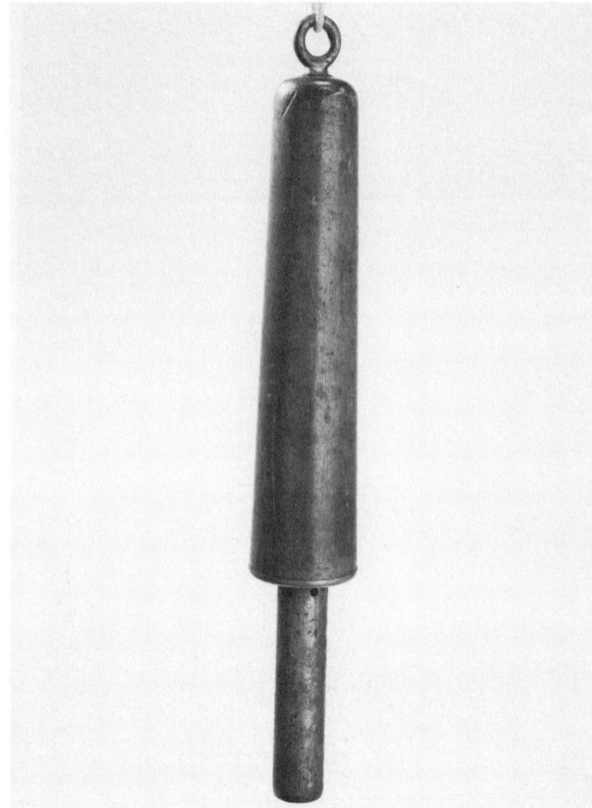

Figure 114. Challenger's tube lead.

served, according to the nature of the bottom (figures 114 and 115), while elsewhere the *Hydra* sounder was used (see figure 56), loaded with the necessary number of iron sinkers. When the new Baillie sounder became available, twelve of these were sent out to the *Challenger* in 1873. These were sturdier than the *Hydra* sounders, with a different detaching mechanism (figure 116), and they proved capable of sounding on every occasion that they were used, though in the deepest water this was a slow task. Keeping the ship over the point where the sounder had entered the water, the line was run out, noting the time that each fathom-mark passed. This time interval increased regularly with depth, until a sudden lengthening of the intervals showed that the sounder was no longer dragging the line down. With 3 cwt of sinkers it could take well over one hour for the Baillie sounder to reach bottom in 4000 fathoms.

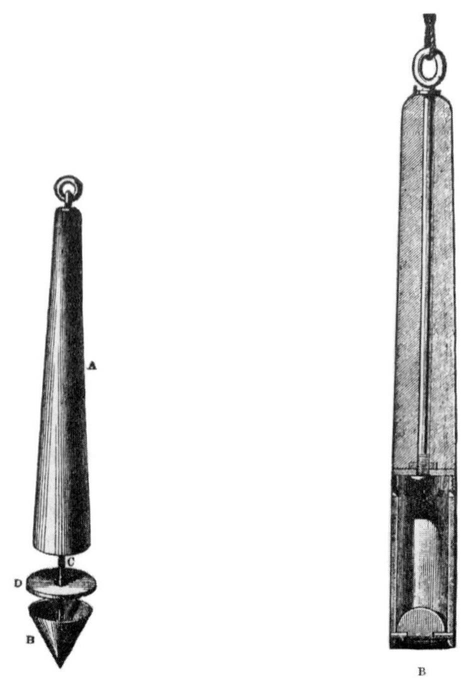

Figure 115. The cup lead and tube lead used on *Challenger*. From *Challenger Report Narrative* 1885 vol. 1 Pt 1, Figs 17 and 18.

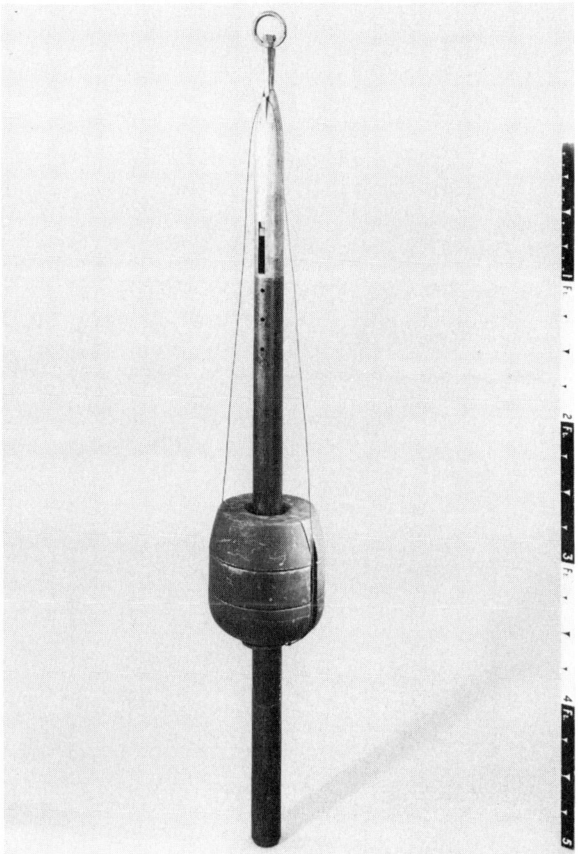

Figure 116. Baillie sounder.

The slip bottle referred to by Wyville Thomson in his account of the *Porcupine* cruise[2] had been invented by Dr Meyer of the Kiel Commission and was used on summer cruises in the North Sea and later on the German research ship *Gazelle* during her circumnavigation of the Earth in 1874–76 (figure 118).[3] The instrument consisted of an open frame down which a cylindrical shell was dropped, being released at the sea bed by a detaching mechanism similar to that of a sounder. With such a bottle there was no danger that water from upper levels would be carried down and mixed with the sample taken at the bottom.

Meyer's bottle was directly adapted into the 'slip' bottle (always referred to by that generic name in the *Narrative*) which served to collect all *Challenger*'s bottom samples. The frame of rods was considered too weak for deep-sea work, and so was redesigned by the Edinburgh brass-founder J Milne, in whose hands it

Figure 117. Model of the midships section of HMS *Challenger* showing the dredging gear rigged for use and a net hanging by the mast.

The *Challenger* carried three sizes of dredging rope, the largest of which was three inches in circumference. That currently in use was racked in the forecastle. Trawling and dredging were carried out from the mainyard with the rope rove through a large accumulator to cushion the strain of the dredge. A platform had been constructed across the central part of the upper deck where the dredge could be emptied. The naturalists were thus able to sieve through the mud and collect their specimens without hampering the sailors working the ship (figure 117).

The waterbottles resembled those in use by other nations at this time; they were non-insulated, fairly small in capacity, and not intended to contain a thermometer. Those for sampling the bottom water differed in form from the mid-water collectors, the first group being known as 'slip' bottles, the others as 'stopcock' bottles.

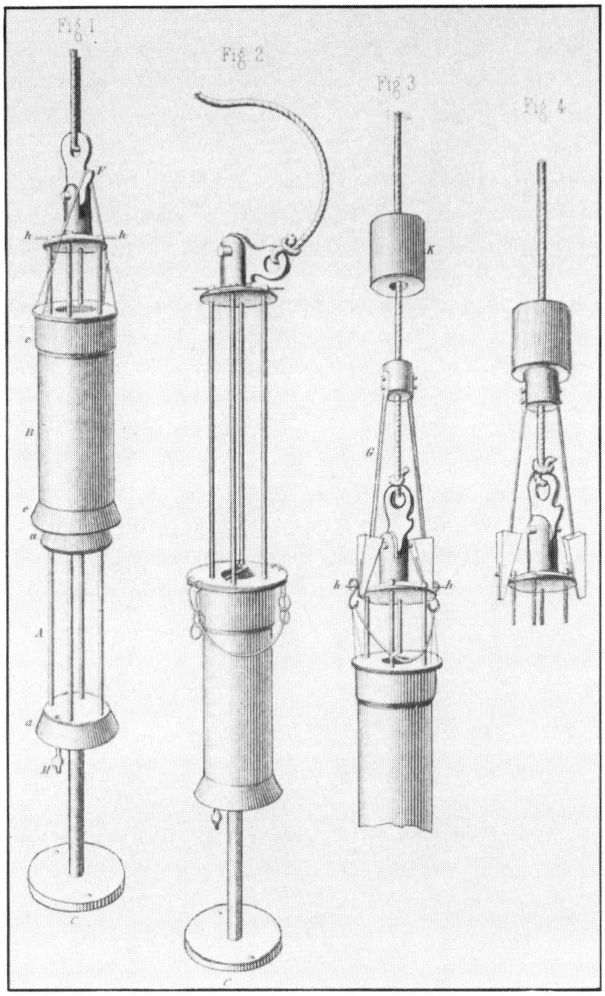

Figure 118. Meyer's waterbottle. Fig. 1, descending; Fig. 2, cylinder detached ready to be lifted; Figs 3 and 4, attachment for mid-water operation by messenger. From *Die Forschungsreise der 'Gazelle' 1872–76* (Berlin) 1889–90 vol. 1, plate 3.

became a solid, all-brass apparatus (figure 119). Efficient closing depended on the shank and end-plates remaining undistorted and, accordingly, Milne

> . . . secured this end in a way which adds equally to the elegance and to the strength of the instrument. The shank and valves are one solid brass casting of the shape shown in the figure, the cylinder is another, and the slipping arrangement E, fixed to the end of the rod F of suitable length and great stoutness, is screwed into the top of the shank, the screw being secured by a rivet.[4]

An arrangement was made to throw off the harness in

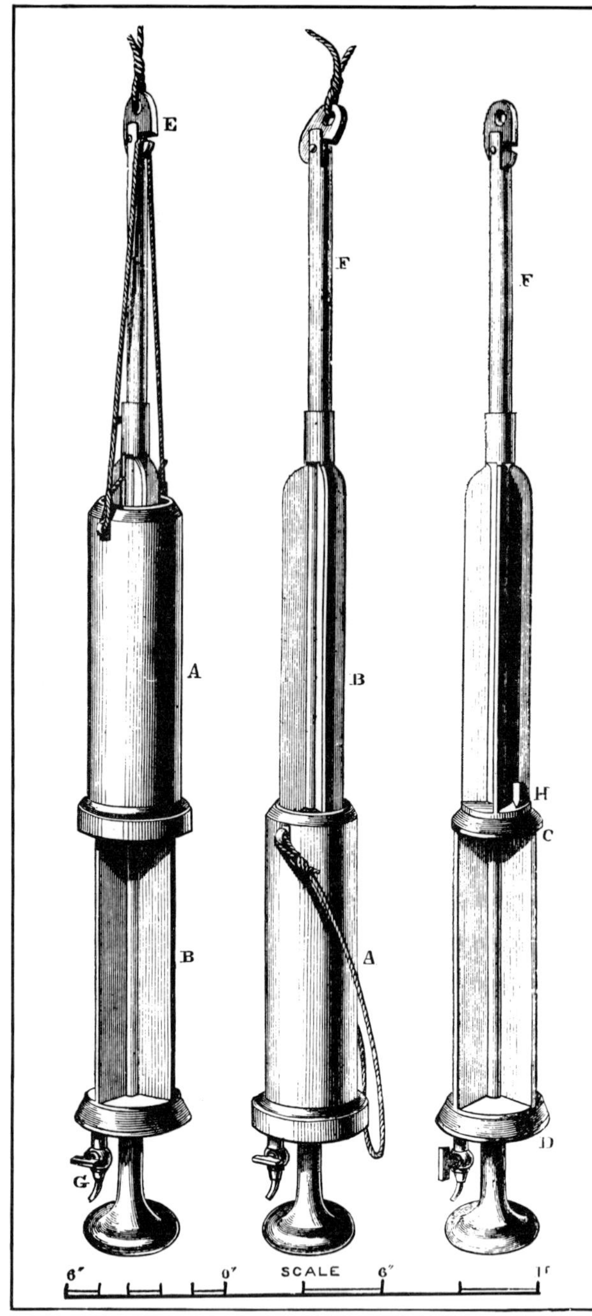

Figure 119. Milne's 'slip' bottle as used on HMS *Challenger*. *Left*, as lowered; *centre*, after the collar harness has been slipped; *right*, cylinder removed to show construction. From *Challenger Report Narrative* 1885 vol. 1, Pt 1, Fig. 39.

mid-water when the apparatus was hauled upwards, although it was never used on *Challenger* in that manner.

H R Mill subsequently adapted this slip bottle for shallow-water use in the Firth of Forth, making it of sheet brass, with a rubber sealing ring on the lower plate.[5] This construction was less than perfect for it was so light in weight that it leaked unless weighted by additional heavy brass messengers dropped down the line to press the cylinder home (figure 120).

The advantage of the slip bottles was their instant operation, in contrast with valved bottles which required lifting through the water for a certain distance before their closing was effected. The slip bottle did however have a tendency to leak on the way up, and so Mill worked to perfect it, to make it more serviceable in deeper water, and, apparently in ignorance of similar developments in progress in Scandinavia (see §10.5), continued to rely on the Milne version of a bottle with a central guide. The improved form was publicised in 1886 (figure 121), by which date the bottle had been in use for a year.[6]

Mill had tackled the problem of leaks in two ways: he fitted rubber washers at the top and base of the cylinder, and provided spring clips which slipped over catches on the lower plate. This bottle, with its positive closing which could be felt by hand on the line through 100 fathoms of water, could be adapted for mid-water sampling and was relatively cheap and easy to construct. Mill found that several, with Scottish pattern reversing thermometers, could be fastened to a single line.

For mid-water samples collected on *Challenger* it remained to find a reliable way of closing the bottle at a selected depth. This was achieved by replacing flap or conical valves by stopcocks whose turning was more positive. However, it was still necessary to find a means of turning the stopcock and for this purpose a flap was provided, attached to an external framework. *Challenger* had been provided with

> . . . a waterbottle of peculiar and ingenious construction, used by Jacobsen in the North Sea Expedition in *Pomerania* in 1872 . . . but it was unfortunately mislaid at the fitting out, and notwithstanding repeated searches was not found till the ship returned.[7]

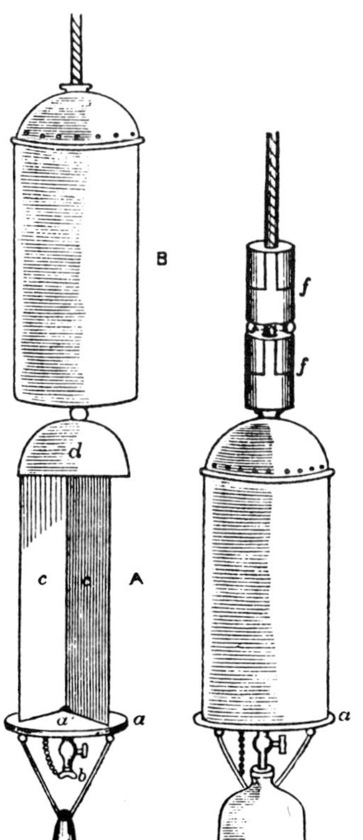

Figure 120. Mill's waterbottle, shallow-water pattern. From Mill 1884 *Proc. R. Soc. Edinburgh* **13,** p32, Figs 1 and 2.

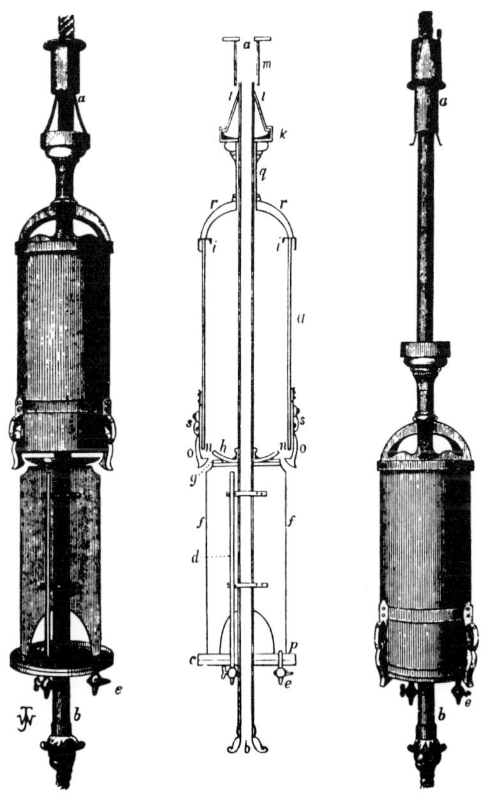

Figure 121. Mill's waterbottle, deep-water pattern. From Mill 1884 *Proc. R. Soc. Edinburgh* **13,** plate 20.

This was probably the small rubber single-stopcock bottle which Jacobsen intended for collecting gases from sea water (figure 122).[8]

The bottle which was used exclusively on *Challenger* for mid-water sampling was designed by the expedition's chemist, John Buchanan. Its correct operation depended on careful handling and Buchanan himself supervised its use on each occasion (figure 123). The stopcocks were closed by a double haul-and-release action before fast hauling up commenced. It took several fathoms of movement to close and thus was not suitable in shallow water or where water-mass boundaries lay close together. Back at home, Buchanan made several modifications to the bottle to make it more reliable, and he subsequently used it on cable company ships and on his voyages with Prince Albert of Monaco, for whom he worked in later years (see §10.6).

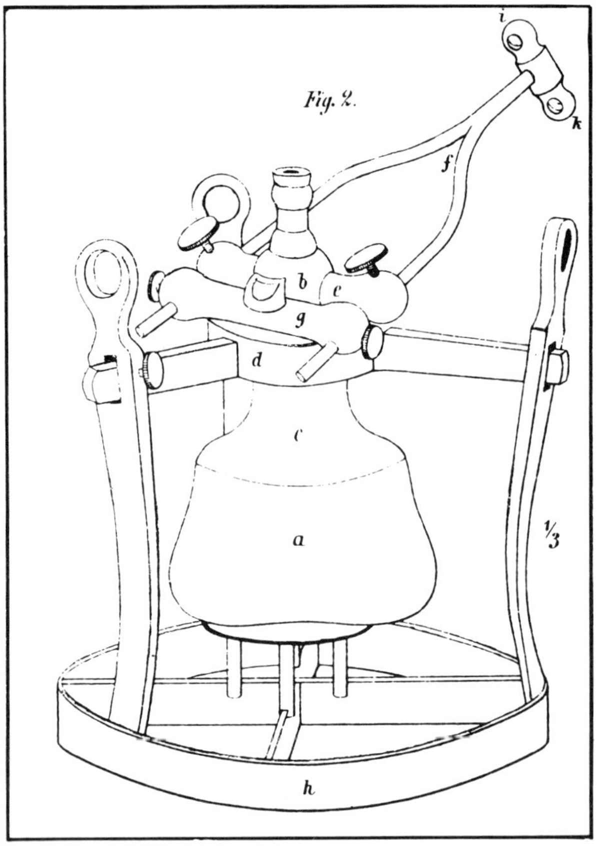

Figure 122. Jacobsen's waterbottle. a, rubber bag; b, stopcock housing; c, collar holding the bag; d,h, frame; e, stopcock turned by lever f, counterbalanced by weight g. From Jacobsen 1873 *Ann. Chem. Pharm.* **167** plate 1, Fig. 2.

The thermometer used for almost all the observations made from *Challenger* was the Miller–Casella model, fitted in a pierced copper case (see §8.4). Several were usually lashed a fixed distance apart on the line. In temperate and tropical regions a temperature profile could be drawn from this 'serial' sounding but her scientists encountered great difficulties when, in 1874, the ship entered Antarctic waters. The first observing station in the region was close to a large tabular iceberg, judged to extend about 1800 feet beneath the water, which was melting rapidly. The temperature sounding was unsatisfactory, all those thermometers sent below 50 fathoms returned showing the same result.[9]

Three days later another serial sounding was made. This time some of the brash ice surrounding the ship was collected and the thermometers cooled to 30.2 °F on both maximum and minimum sides before immersion. From 50, 100 and 200 fathoms they came up registering 30.2 °F on the maximum scale and 29 °F on the minimum, showing, in the opinion of those concerned, that temperature was uniform at those depths. Those sent to 300, 500 fathoms and to the bottom (1675 fathoms) gave different results, the first two registering 32.8 °F maximum and 29 °F minimum, while the lowest instrument registered 33 °F maximum and 28.8 °F minimum.[10]

It appeared that temperature increased from 29 °F at 280 fathoms to 32.8 °F at about 500 fathoms, but it was impossible to say what happened below that depth, other than to state that at the sea bed it was less than 33 °F and more than 28.8 °F. The layer of cold water appeared to be a local phenomenon, related to the depth of nearby icebergs, and remaining at that level because the greater specific gravity due to lower temperature was more than offset by the lower salinity of melting ice.

The next station yielded much the same inconclusive result. No temperatures were taken at the following one and at the final station in Antarctic water, adjacent to Termination Land, the scientists were again frustrated. They obtained readings down to 150 fathoms, where the temperature was 34 °F, but from here to the bottom, 1975 fathoms, they could only state that the figure lay somewhere between 34 and 31.9 °F. Their uncertainty was partly due to their having omitted to note the original position of the indexes. The thermometer lowered to the bottom showed 31.3 °F minimum but it might well have been cooled to this point before being let down.[11]

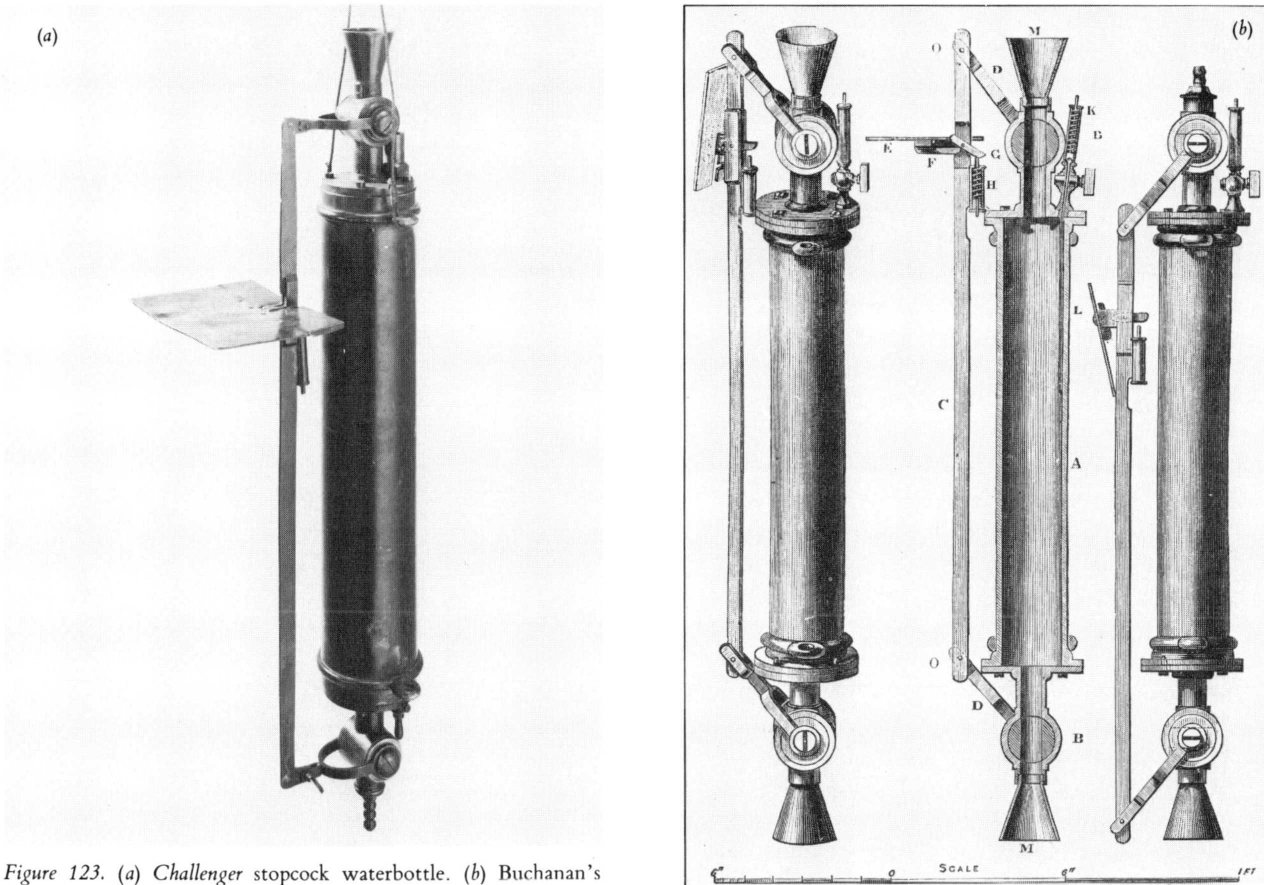

Figure 123. (*a*) *Challenger* stopcock waterbottle. (*b*) Buchanan's stopcock waterbottle as used on *Challenger. Left*, open as during ascent; *centre*, sectional drawing showing the bottle in the act of closing; *right*, closed and ready for hauling up. From *Challenger Report Narrative* 1885 vol. 1, Pt 1, Fig. 41.

Where small localised wedges of water of markedly differing character were present, *Challenger*'s widely spaced soundings were very difficult to interpret. This led to some discrepancy between figures given in the *Narrative* and those in the tabulated results. The *Narrative* figures have been quoted here, for they were published later and represent the *interpreted figure,* not necessarily that indicated by the thermometers.

9.3. LATE ARRIVALS: THE BAILLIE SOUNDER AND THE REVERSING THERMOMETER

As mentioned above, examples of the Baillie sounder were sent out to the *Challenger* during her voyage. This apparatus was invented by Lt Baillie in 1873 (figure 116). It was similar to the *Hydra* sounder but was considerably heavier. Though the detaching action took place at the sea bed the tube penetrated well into soft sediments and recovered a lengthy core. A butterfly valve had to be fitted when working over unconsolidated material but on a stiff clay bottom this was unnecessary and an undisturbed core ent ered the tube. The tube lead, used in water less than 1000 fathoms, was simply a short detachable section of tube, identical to that of the Baillie sounder, screwed to the base of a deep-sea lead (see figures 114 and 115, *right*).

The important and novel reversing thermometer (see §8.5) was sent out in 1874 and first tried in the shallow water of the Sulu Sea in the autumn of the same year. Trouble with the original reversing frame was corrected and the thermometer tried again the following February and from time to time thereafter. Additional thermometers were sent out from England but several were broken in transit and only four were used regularly. In all, some 48 observations were made between June 1875 and March 1876, at depths down to 2775 fathoms. By comparison with the Miller–Casella thermometers, 30 of the readings were higher, three were lower and 15 similar. This variance was believed to arise from certain design defects. If, for example, the mercury thread broke slightly above the constriction, the reading was too low, and it was assumed that this had been the cause of the three lower readings. It was less easy to explain the higher readings; perhaps the constriction in the tube had been closed by pressure at a higher and therefore warmer level, or maybe compression had been transmitted through the mercury-filled sheath. If so, the temperature registered would not be that at the level of reversing. Indeed, both causes might operate concurrently. All the reversing thermometers tested on *Challenger* were of the U-form, requiring a 360° turn for their operation.

During *Challenger*'s voyage her chemist Buchanan constructed two forms of piezometer to check both depth and temperature as measured by the conventional sounding line and Miller–Casella thermometer (figure 124).[12] Buchanan took as his starting point the fact that water is changed in volume more by pressure than by temperature (especially the near-freezing temperatures of the deep oceans), whereas mercury is changed in volume more by temperature than by pressure. In his water piezometer, compression acted on the glass and the mercury since one section of the tube consisted of a flexible rubber membrane. The piezometer was sent down together with a thermometer and the difference between their readings was taken to indicate pressure.

The mercury piezometer resembled a Miller–Casella thermometer and in fact trial examples were made from such thermometers with the tube opened to allow pressure to act on the contained liquids. Mercury filled both ends of the piezometer tube, around a registering sector of water. Several forms of these instruments were made, but none of them seems to have been entirely successful. Buchanan made four mercury piezometers on board *Challenger* before one lasted long enough to yield a series of results. This he used 'as often as possible' between Tahiti and Valparaiso. After the voyage he went on to develop a pressure sounder which he patented,[13] but it seems not to have been used and he was overtaken by the course of events. By the time that *Challenger* returned the reversing thermometer and wire-sounding machines were in common use for deep-sea surveys and were giving good results.

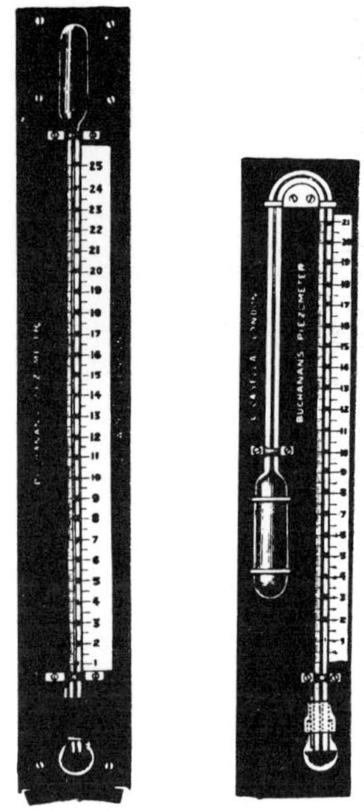

Figure 124. Buchanan's piezometers. *Left*, the water piezometer; *right*, the mercury piezometer. From Wilkinson 1896 *Submarine Cable Laying and Repairing* Fig. 189.

9.4. TAIT RE-EXAMINES *CHALLENGER*'S THERMOMETERS

After *Challenger*'s return 30 of her Miller–Casella thermometers were retested by Peter Tait, a former Belfast colleague of Wyville Thomson and later Professor of Natural Philosophy at Edinburgh University. It was suspected that the pressure correction applied to the observed readings was excessive.[14] Tait reported that on inspection he found the thermometers rather dissimilar; the size of each degree varied and the indexes moved erratically. Such 'unscientific' instruments were, he admitted, probably better suited to rough handling than the delicate conventional thermometers used ashore. When testing the thermometers originally, Davis and Casella had gone to considerable lengths to ensure that the compression chamber was cooled before any measurements were made. However, Tait found that the ebonite, chosen as a backing in preference to wood for its resistance to warping and waterlogging, had the property of retaining heat to the detriment of its reading. He also found that compression of the mercury contained in the thin-walled aneurisms at the bend of the tube affected the readings.

After long and painstaking experiments, Tait concluded that Davis had relied on laboratory techniques and that his corrections were inapplicable to a thermometer within the deep sea. Faults arose from a pressure effect on the unshielded parts of the tube; the heat of the test tank water which had varied from day to day; the quality of ebonite as a heat retainer, and the uneven pressure gradient operating across the glass forming the outer pressure sheath, which might induce a temperature change in the glass itself. Since, with the exception of the first, these faults occurred only under test, that alone should have been taken into account. Hence, Davis's table of corrections was indeed excessive.

Regrettably, Tait's work was largely in vain. The Miller–Casella thermometers issued to *Challenger* had, as Tait himself noted, very uneven graduations on their scales. They covered a large temperature range so that the space allotted to each degree was short, varying between 1.2 and 1.5 mm. The breadth of the engraved lines marking degrees was 0.12 mm, so that it was not possible to read the instruments with confidence to less than 0.5 °F.

This leads to a consideration of the report itself. *Challenger*'s voyage is celebrated as a great event in British, or even world, oceanography. The many volumes of the report, published between 1880 and 1895, are outstanding in their careful descriptions and beautiful illustrations, mainly in the field of zoology. Yet the ship carried apparatus more antiquated than that in current use elsewhere and, as Tait showed, the accuracy of some instruments left much to be desired.

Looking at the *Narrative*, in particular Volume I, Part 1, in which most of the details concerning instruments and apparatus are given, we find first a historical resumé seen from the British viewpoint, written by H R Mill. This was followed by a comprehensive account of the apparatus that was actually shipped on *Challenger*. Three new instruments put on board—a Thomson sounding machine, Jacobsen's waterbottle and a Siemens electrical thermometer—were either never used or discarded after a short time. Embedded in the day-to-day narrative of events are accounts of instruments evolved *after* the return of the *Challenger*. Several of the illustrations refer to these later developments and it is obvious from quoted reports such as that on the Siemens thermometer, that this is not just a report of the voyage of the *Challenger sensu stricto*, but a continuing history of oceanography right up to 1885, the date of publication.

Much of Buchanan's work arose directly from his experiences on board *Challenger*, but this may have led people to give more credit to the physical side of the findings than they merit. *Challenger* set boundaries to the physical parameters of the ocean, delineating the general temperatures and specific gravities, the average depths of water on the continental shelves and down into the abyss. With observing stations hundreds of miles apart no more could be expected. It was up to later expeditions to research intensively within limited areas, returning to the same stations every year to observe the fine detail and to note the seasonal and annual variations, and it was on such cruises that the insulated and reversing waterbottles, the accurate thermometers and the current meters described in the following two chapters, came to be developed.

REFERENCES AND NOTES

1. *Challenger Report Narrative* 1885 vol. 1 Pt 1, pp34–5
2. Wyville Thomson C 1873 *The Depths of the Sea* (London) p500
3. The full report of the expedition is *SMS Gazelle: Die Forschungsreise 1874–76* (5 vols) (Berlin) 1889–90. Notes on the apparatus used are given in vol. 1
4. *Challenger Report* 1885 *Op. cit.* pp111–12
5. Mill H R 1884–5 'On the salinity of water in the Firth of Forth' *Proc. R. Soc. Edinburgh* **13** pp31–3
6. Mill H R 1885–6 'On waterbottles, with the description of a new form of slip waterbottle' *Proc. R. Soc. Edinburgh* **13** pp539–46
7. *Challenger Report* 1885 *Op. cit.* p117
8. Jacobsen O 1873 'Über die Luft des Meerwassers' *Ann.*

Chem. Pharm. **167** pp1–38
9. *Challenger Report* 1885 *Op. cit.* p398
10. *Challenger Report* 1885 *Op. cit.* p399
11. *Challenger Report* 1885 *Op. cit.* p409
12. *Challenger Report* 1885 *Op. cit.* pp102–8
13. *GB Patent Specification* No. 2472 (1879)
14. Tait P G 1881 *Challenger Report Narrative* vol. II Appendix

Chapter 10

Hydrographic Research for Fisheries Conservation

10.1. EUROPEAN COOPERATION IN FISHERIES RESEARCH

Before the middle of the nineteenth century it was popularly said that 'there were more fish in the sea than ever came out of it'. In later years this was clearly not so, as far as European fishermen were concerned. Steam-powered drifters and trawlers went to sea independent of wind and tide; the newly invented otter trawl scraped many spawning-grounds bare of eggs and young fish, while for reasons not understood the herring deserted many of their regular feeding grounds.

Economic collapse of the fisheries and the consequent hardship suffered by the traditional fishing communities of northern Europe eventually forced governments to set up committees of inquiry. But fish are no respecters of territory, and often the depleted spawning grounds lay in the offshore waters of one nation while another felt most severely the loss of mature fish. Stocks could be conserved only by international agreement on the delineation of non-fishing zones and on the sizes of net mesh and types of gear that should be permitted. Limited research cooperation existed by the 1890s: Norway was concerned with the cod-fishing grounds of the Arctic and North Atlantic, English and Scottish scientists were exploring North Atlantic and North Sea waters, while Swedish and Danish investigations were largely confined to the Baltic Sea. In 1892 Professor Pettersson of Stockholm proposed an international hydrographic survey of the North Atlantic, North Sea and Baltic in which he outlined the research topics and stressed the need for repeated observations at fixed stations.[1] This proposal was repeated at the Sixth International Geographical Congress held in London in 1895.[2]

H N Dickson, one of Britain's participating scientists, described the attempts by the crew of HMS *Jackal* to carry out the work specified, but the voyages were severely hampered by bad weather, poor equipment and a very unsuitable ship.[3] In 1893 only one waterbottle was carried, a Mill self-locking type (see §9.2) and this could not be risked in deep or rough water until the strength of the sounding line was assured. In 1894 two bottles were taken but their closing mechanism was thought to be uncertain. Hemp line was still used for sounding and the results were generally in excess of check tests made with a Thomson sounding machine and pressure tube sounder. One of the problems was that suspect data could be compared only with results from the cruises of *Porcupine*, *Knight Errant* and *Triton*, gathered up to 25 years previously and from stations some miles distant. There was always doubt as to whether conflicting figures represented instrumental errors or genuine changes in conditions over time or place.

The numbers and distribution of fish species inhabiting a particular region are controlled by average or mean sea conditions, and they fluctuate as conditions depart from that mean. As Pettersson stressed, the only way to understand how conditions changed was to repeat observations, time and again, season by season, year by year, so that mean salinity, temperature and current strengths could be established. Given the terrible weather conditions that sweep through the

Arctic and North Atlantic during the winter, this plan was easier to state than to execute; nevertheless, in time it was undertaken in large part.

Annual variations in the weather have considerable effects on the shallow seas of north-western Europe. The persistence of the wind from one quarter, heavy rain leading to excessive river discharge, or years of drought, are all reflected in the salinity and temperature of the water and the degree of mixing that takes place through the water column. Even water depth and sediment type may be altered by man's activities. Dredging for sand and gravel, construction of harbours or offshore buildings, adoption of new types of fishing gear, all vary the patterns of deposition, erosion and scour across the sea bed upon which many species of fish lay their eggs. Scandinavian nations were among the first to work intensively towards both a practical and a theoretical understanding of the interaction of weather and sea conditions. In their hydrographic work, the name of Fridtjof Nansen (1861–1930) stands out as a promoter of the most rigorous methods of investigation.

Nansen's formal training was as a zoologist but as a boy he delighted in the conquest of the ice and snow of the Norwegian winters and soon set his sights on Arctic travel. He was of extremely practical bent and was always building and testing novel apparatus for his expeditions, while his methodical planning contributed in large measure to the survival of any team venturing into the isolated and dangerous regions that he yearned to explore. This combination of physical strength, a practical nature and attention to detail brought success when he set out to unravel the complexities of the northern waters. From 1882 he was engaged in zoological research at the Bergen Natural History Museum. After receiving his doctorate he headed the first expedition to cross the Greenland ice cap in 1888 and went on to lead the Norwegian North Polar expedition in 1893–96.

The North Polar expedition was a bold attempt to demonstrate the existence of a trans-polar current moving across the Arctic Ocean to emerge in the Norwegian Sea. Most of this ocean is covered with ice, and so to prove that this ice is drifted by the underlying current Nansen proposed to let his ship become frozen into the ice pack and be carried across the pole to be released in the region of Spitzbergen (now Svalbard). He put his plan before a meeting of the Royal Geographical Society in London, where it was attentively received by many of the Arctic explorers of the day and by hydrographers, but with little encouragement.[4]

Nansen's ship *Fram* was built with a strong, rounded hull that would suffer no damage when gripped by the pack ice. During the voyage (if one may call it that) Nansen, seeing that *Fram* would not pass directly over the pole, left the ship and with a few companions attempted to reach it by sledge. In this he was unsuccessful, but his theory was confirmed for *Fram* did indeed drift through the ocean just as Nansen had foretold. Throughout her passage, the crew measured water temperatures and salinity, and the report of this and all their other work filled six large volumes published soon after their return.[5]

Nansen's daring exploits, and the return of his ship and crew after the years of silence when many believed them lost, overshadowed the very real contribution to Arctic hydrography and zoology which the report demonstrates. Several biographies of Nansen were written at this stage of his career but his later oceanographic work has not been fully documented. On his return to Norway he became Professor of Zoology at the Royal Frederick University and he played an active political role when the united kingdoms of Norway and Sweden were separated in 1905. He was Norwegian ambassador to Britain between 1906 and 1907, returning to the Royal Frederick University as Professor of Oceanography in 1908, and he went on cruises in the North Atlantic in the prewar years 1910–14. In addition he was appointed director of the ICES Laboratory when that body was founded in 1902 (see below) and it is his work there that set the standards of precision of both instruments and method that were to become generally accepted in fisheries hydrography.

While *Fram* was on her epic voyage, the Danish *Ingolf* expedition spent the summers of 1895 and 1896 exploring deep waters off the coasts of Iceland and Greenland. On board was Martin Knudsen (1871–1949), a young physicist who later joined the staff of ICES and there developed the methods used to test and analyse the sea water samples received.

Ingolf, which had zoological aims, carried apparatus that had been proven on board the American ship *Blake* and also on the German 'Plankton' expedition.[6] For shallow under-way soundings a Rung pressure sounder was deployed from a Thomson machine, for deep soundings she was equipped with a chain-driven Sigsbee sounding machine. Sigsbee waterbottles with a capacity of 1245 cc collected water samples. On the first voyage temperature measurements were made with Negretti and Zambra reversing thermometers, fitted in Magnaghi reversing frames, but Knudsen found that, when hoisted through turbulent surface water, their readings were ruined by additional mercury escaping from the bulb. To prevent this, he designed a version in which the bulb was folded back upon itself (figure 125) and the French instrument-maker Chabaud prepared some examples for him to test. However, due to poor manufacture these too proved erratic and finally Knudsen had others made to his pattern by Negretti and declared himself satisfied with their performance during the second voyage. This so-called 'Knudsen modification' of the reversing thermometer was made commercially by Negretti and Zambra for some years, and it served on various expeditions, until it was eventually replaced by an even better design made by the German firm of Richter.[7]

The third series of cruises operating in Norwegian waters was that undertaken by Johan Hjørt. In his introduction to the *Report of the Norwegian Fishery and Marine Investigations* covering the years 1897–99,[8] Hjørt explained that these had originally been set up firstly to solve the question of the influence of ocean currents on fisheries by studying hydrography and pelagic organisms, and secondly to study the distribution of young fish stocks around the Norwegian coasts. A paper by Hjørt and Gran in the same volume showed how much had been accomplished in these first years.[9] Deep-lying currents had proved very difficult to measure by direct registration, and hence a theoretical method had been borrowed from Pettersson and Gustaf Ekman (working in Swedish waters) in order to determine the salinity and temperature of each layer of water, since it had been recognised that such layers preserved their identity over long distances. It had for example been found that all water in the Skagerrak with salinity less than 30‰ came via the Baltic current; all that over 32‰ came from the North Sea and Atlantic. Deep water with salinity over 35‰ came from the Gulf Stream as an undercurrent through the Norwegian Depression; other water of 34–35‰ being also found to originate from the North Sea. By taking careful measurements of salinity, therefore, the hydrographic character of the sea could be comprehended at any time and it was possible to follow changes throughout the year as the different currents increased and decreased their relative strengths. The Skagerrak changed its character markedly twice each year and this apparently influenced the seasons and therefore the locations at which herring were to be found off the coast. The fish were repelled by the increased depth of fresher Baltic Sea water during the summer months. If irregularities in the year-to-year catches of herring could be thus explained, this would be of great value. A similar derivation of waters could be found by examining the plankton contained. The seasonal variations were found to be clearly affected by the strength and direction of the prevailing winds.

Encouraged by these findings, the Norwegian government commissioned a new ship, the *Michael Sars* (named after a pioneer marine zoologist), which made its first cruise in 1900. This ship was 125 feet in length, 23 feet in the beam, with a depth of 12 feet. Her 300 hp engines drove her at 10 knots and she carried sails purely to steady her while working. All research ships must choose between the ability to carry out intensive measurements within a restricted area, or wide-ranging work across the world's oceans. The *Michael Sars* was built with a restricted area specifically in mind and so she carried all the gear normally fitted on a Norwegian

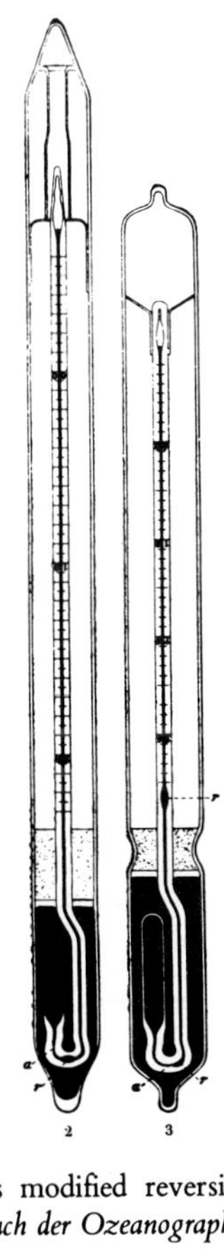

Figure 125. Knudsen's modified reversing thermometers. From Krümmel 1907 *Handbuch der Ozeanographie* Fig. 51.

fishing vessel. But it was possible, by international cooperation, to carry out short research projects for the benefit of the fishing industry, while at the same time collecting hydrographic data for long-term projects of a more purely scientific nature. The ship was provided with a Lucas sounding machine and a deep-sea winch carrying 4000 m of 3 mm steel line, winding in at 120–150 m per minute. Samples could therefore be taken from 1000 m within half an hour.

With the rapid publication of the results of these and other cruises, the value of international cooperation was brought home to various governments, and after considerable political discussion the International Commission for the Exploration of the Sea (ICES) was set up in 1902, with its headquarters at Copenhagen. Its membership varied from time to time, as political feeling decreed, but most of the Atlantic nations from Portugal to Scandinavia did participate in some way and cooperation extended across the North Atlantic to Canada and the United States of America. The governments of these participating nations agreed to contribute staff and funds for its upkeep. ICES was to coordinate the necessary research programmes and to persuade governments to legislate to conserve fish stocks.

The history of ICES has recently been chronicled by one of its past presidents, Arthur Went.[10] The major share of the research effort was devoted to investigations into the life cycles of plankton and fish species but there was also seen to be a need for a hydrographic service to look into the physical parameters of the research areas. The outlines of its programme were first defined as far back as 1899 and at the inaugural meeting of the council in 1902 plans were ready and staff could be appointed to start work. The first assistant to the council, and head of the hydrographic service, was Martin Knudsen. International cruises were to take place quarterly, and the first of these began in August 1902, with ships from Denmark, Finland, Germany, Norway, Russia, Scotland and Sweden taking part. The initial programme looked for conformity of method so that results from the different vessels would be comparable. The results were published by August of the following year and a similar pattern of operations continued until World War I.

Under the control of ICES an International Laboratory was set up at Christiania, now the city of Oslo. This site had two advantages over that of Copenhagen, where the council headquarters were housed. Firstly, it had access to deep water necessary for testing new apparatus; secondly, it enabled the council to secure the (voluntary) services of Nansen as Director throughout its existence, from 1902 to 1908. It is clear from the reports of the laboratory that Nansen did not regard this duty as a sinecure, but was closely involved in all aspects of its work and in its funding and management. The laboratory started work in September 1902 with V Walfrid Ekman as first assistant, but when Nansen became involved with political affairs at the time of Norway's independence, Ekman was promoted to sub-director. Instruments were developed to meet specific needs, but even with restricted research aims it was never possible to make one single type of bottle or current meter that was ideal in all waters. Considerable effort went into devising methods of use, for which purpose *Michael Sars* often served as test ship.

The era of fisheries research was marked by a change from long voyages to distant waters, to a series of briefer local cruises. Partly for this reason, and partly due to closer contact between research workers, instruments were developed, tested and reported on far more quickly than had formerly been the case. A particular type of waterbottle, for example, might be used on three research ships simultaneously; by the end of that season each group would have modified the instrument to suit the characteristics of the area in which it had worked. Instead of tracing the progress of each series of voyages it is therefore useful to look at the evolution of the instruments as a continuing history, noting their use on a particular ship where this is well documented.

During this period oceanography was not restricted to the practical gathering of measurements; as data built up, theories of the dynamic structure of each ocean basin or sea area were formulated and tested against observed conditions. Nansen himself had such dynamic concepts in mind when he postulated the existence of an Arctic Ocean gyre which would carry *Fram* through that ocean. The later work of Norwegian meterologists and physicists, Ekman among them, became justly renowned, laying the basis for our present-day understanding of oceanic circulation and the interaction between the atmosphere and ocean.

ICES continues to the present day, but this history is brought to a close in 1914, a year which in so many ways marked the end of an era. It was to be succeeded by one in which increasing use was made of underwater acoustics, initially to detect the sea bed and later the positions of certain water-mass boundaries. But before then the instrusion of such military machines as submarines and torpedoes into the deep seas, coupled with a political situation whereby the northern waters became a theatre of war, led to much gathered information being restricted, while the focus of research was directed along new lines.

10.2. NEW MATERIALS FOR INSULATED WATERBOTTLES

By the second half of the nineteenth century new materials with good insulating qualities were being manufactured to replace the porous woods, wool or canvas which relied on air-filled pockets to reduce thermal conductivity, but which lost their insulating properties when submerged in the sea.

The first of these new materials, rubber, had long been familiar, and had been traded from tropical America before the arrival of the Spaniards in the sixteenth century. However, its elastic properties in the semi-treated state were ephemeral and were destroyed by extremes of temperature. During the eighteenth century French scientists began working on it and by the beginning of the nineteenth century its chemistry and solvents were well known. Over the next 40 years the rapid advance of pure and applied science combined with the practical discoveries of a few men to lay the foundations of a great industry based on the unique properties of rubber.

Rubber sheet, tube and belting were the first items to be manufactured, and by 1839 the now-standard manufacturing processes had been discovered. Vulcanisation for instance (whereby rubber is heated with sulphur to yield a hard and durable material) was formulated in America around 1833. After 1843 similar experiments were repeated in Britain and a range of hard rubbers, known by trade names such as Vulcanite and Ebonite, was patented. Depending on the proportion of sulphur, and the 'cooking' temperature of the mix, rubber could serve as an insulator, as a watertight seal or as a durable elastic. As noted in Chapter 8, it was made into thermometer backs and it was also used to insulate thermometers.[11] It was ideal for sealing waterbottles, while strong elastic bands fitted in accumulators (illustrated in figures 57, 93, 113 and 117) eased the strain on sounding and dredging lines, thus reducing the risk of breakages.

Some gutta-percha, a coagulated sap drawn from several varieties of trees grown in Malaya, was first sent to Britain by William Montgomerie and its properties were brought to the attention of the Royal Society of Arts in 1843. Faraday examined samples and thought that it might be useful as an electrical insulator. In 1845 a machine was devised to extrude it into tubes and hoses, and from 1848 it was used to cover wires. Its main service was undoubtedly to the rapidly expanding telegraph industry where it insulated many thousands of miles of telegraph cable, both inland and under water. Its qualities as an electrical insulator were affected by temperature and pressure, but these factors could be taken into account during cable manufacture. The material was also a good thermal insulator and so was used on some of the early insulated waterbottles and thermometers.

Celluloid, originally the trade name of a thermoplastic composed of cellulose nitrate and camphor, was invented by Parkes in 1862. Small quantities were made but, being highly flammable, it was replaced by cellulose acetate after 1894. Easy to work and mould, it commonly replaced bone and ivory for small items of domestic and ornamental use.

Bakelite was the trade name of various synthetic resins. These are thermosetting, having the property that once heated and worked, they set and cannot be softened and reworked by additional heating. Both celluloid and Bakelite conduct heat poorly and have the virtue of being light in weight. Rubber and gutta-percha were subject to adiabatic heating under compression. As a smaller volume of celluloid or Bakelite gave the same degree of insulation, its use reduced this potential source of error.

Metals too emerged from the chemistry laboratory in the late nineteenth century. Aluminium was isolated in 1845 and around 1854 was being made in small quantities as a curiosity. After 1877 some was prepared commercially and ten years later it was being produced electrolytically in Switzerland, American production soon followed, and the British Aluminium Company began its manufacture in 1896. Its main advantage was of course its light weight, relative to other metals, but for oceanographic apparatus this was not at first an advantage. However, when sensitive current meters were being designed such light materials were essential; for example, aluminium was used for the fins of some of the meters described in Chapter 11. The hard-wearing or non-magnetic properties of other alloys led to their use as bearings, as non-magnetic metals (essential for instruments containing a compass) and for their resistance to corrosion. We have already seen that Sigsbee's waterbottle had bearings of German silver, an alloy made up of five parts of copper, two of zinc and two of nickel. Phosphor bronze, a tough alloy containing up to 2.5% phosphorus, was preferred to conventional lead shot in Ekman current meters, for small bronze shot could be punch-marked for identification whereas this was not possible with soft lead shot.

10.3. INSULATED WATERBOTTLES OF EKMAN (1877) AND PETTERSSON (1890)

Frederik Laurents Ekman (1830–90) was a chemist by training and by profession, who taught at various colleges in Uppsala and Stockholm. He assisted with the Swedish Academy's researches in the Baltic and neighbouring seas and for this work he designed a series of waterbottles, some of which he later showed at various exhibitions and congresses. His first bottles were not insulated since his aim was to collect water from specific levels for gas and salinity analysis. Subsequently he experimented with various insulating materials so that he could obtain temperature measurements of the bottle's contents.

The history of the development of Ekman's bottles was finally set down after his death by Pettersson, who also completed the analysis and tabulation of the results of the expeditions to which Ekman had contributed.[12] The numbered sequence of that account is followed here; the earliest non-insulated bottles are described in §10.5.

Ekman's first insulated bottle was based on the design of his No. 3 (see §10.5 and figure 126). He showed this at Philadelphia in 1876 and it was afterwards used at Käringön, off the west coast of Sweden. It was insulated by end-plates of ebonite and a double cylinder wall packed with colophony resin. For the major expedition of 1877, during which the ships *Graf Klint* and *Afhild* sailed 2100 nautical miles and took 1800 temperature measurements at 200 stations, Ekman designed a new and simpler bottle, No. 5 (figure 127). His own description, translated and rephrased, reads:

> A square galvanised iron frame, with four strong horizontal iron ties, braced and protected the other parts from impact. Two solid steel springs under the frame damped out any collision with the sea bed. The cylinder was fixed within the lower part of the frame.
>
> The part which was to form the lid and bottom of the cylinder was a sleeve movable along a fixed central stem and carrying at its upper end a brass flange holding the sleeve up as the instrument descended through the water.[13]

Ekman found that his new design offered better heat insulation. The overall weight of the bottle was greater so that it had to be used from a winch, but the moving parts were less heavy. The bottle closed as soon as its

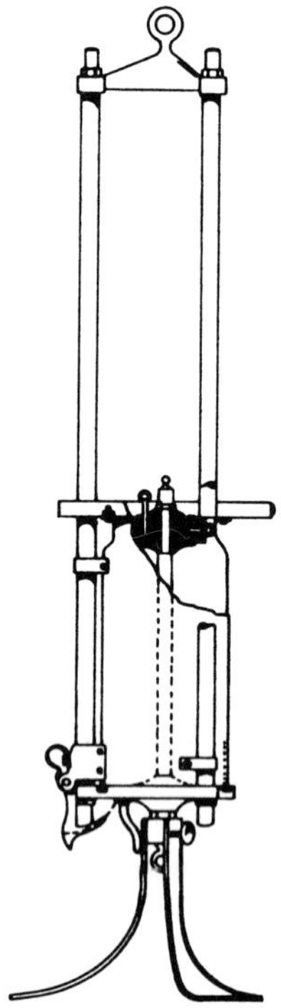

Figure 126. Ekman's insulated bottle, No. 3 design (see §§ 10.3 and 10.5). Redrawn from Ekman and Pettersson 1893 *Kong. Svenska Vet. Akad. Hand.* **25** No. 1 Fig. 5.

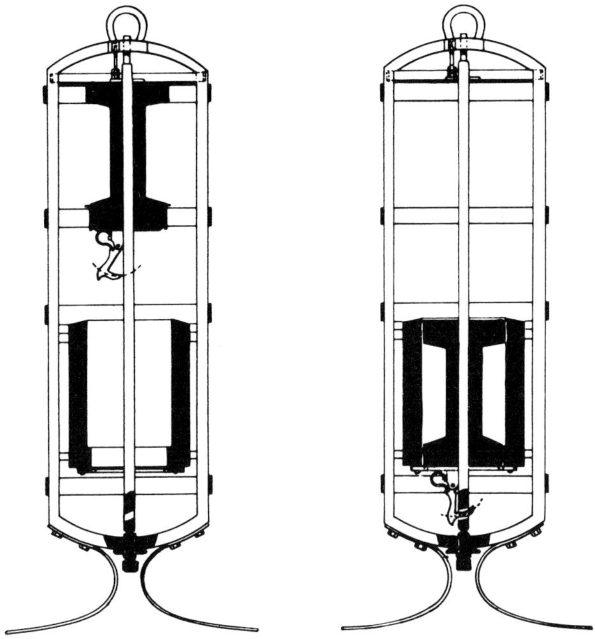

Figure 127. Ekman's insulated waterbottle No. 5. *Left*, open during descent; *right*, closed. Redrawn from Ekman and Pettersson 1893 *Kong. Svenska Vet. Akad. Hand.* **25** No. 1 Fig. 14.

descent was checked, the central portion being clamped in the cylinder by a hook. The base of the sleeve was belled out into a brass disc covered with rubber, which fitted over the lower cylinder orifice. The upper seal was also of rubber. The bottle was clad with 25 mm of gutta-percha and this material, being lighter than sea water, partly offset the weight of the central sleeve as the bottle sank. Back on board, an ebonite plug was removed from the upper cover in order to insert a long-stemmed thermometer.

With this bottle Ekman was able to discover a great layer of cold water existing at depth during the summer months which had escaped the notice of previous observers equipped with Miller–Casella thermometers. The bottle was subsequently used by Buys-Ballot and by Nordenskiold during the *Vega* cruise around parts of the north Russian coast in 1878–80.[14] Meyer took it as the inspiration for his slipping-cylinder bottle and its design was later adapted by Pettersson for his own insulated waterbottle.[14] Pettersson praised its achievements, even though by the time he was writing (1894) it was being carried only as a reserve instrument, being unsuited to working on the wire lines then in common use.[15] The instrument, as described in 1881, was available from the instrument-maker Lundquist of Stockholm, in both insulated and non-insulated forms. A simplified form of Ekman bottle was illustrated by Richard in 1907.[16] The cylinder, rather than the central shaft and lids, dropped to form the container. As in the earlier models, the cladding was 25 mm of gutta-percha, and temperature was measured on return by passing a thermometer through the lid.

As might be expected, Ekman's bottle suffered from the defect of an insecure closing mechanism, being dependent on steady motion down through the water to keep the sliding parts raised. The closing shock could be felt by hand on the sounding line down to 80–100 m only. It also worked best on a hemp line, so that it became something of a liability on vessels equipped with wire handling gear.

Otto Pettersson (1848–1941) was also a chemist who taught first at Uppsala and later at Stockholm, during which time he invented the waterbottle that bears his name. He was active in the negotiations preceding the foundation of ICES and he became its first honorary vice-president in 1902. Pettersson's apparatus clearly owes its form to Ekman's non-insulated bottle (see §10.5) but its insulation was novel. The bottle was in use from about 1890 but the first published description did not appear till 1894 (figure 128).[17] Within the cylinder were three concentric sheet-brass liners, held by ebonite spacers, and closed by lids of metal covered by rubber sheets. The design was such that no metal connection or screws passed through this insulation. A propeller in the upper frame gave a positive closing action. As the bottle was raised the propeller released the upper lid which fell onto the cylinder, overcoming its weak supporting hooks. Assisted by the suspended weight, both parts then slid down to rest on the base plate. Pettersson claimed that

> This arrangement gives perfect protection against changes of temperature in the water without. A thermometer, which is inserted into the inner water column of the apparatus through a hole in the upper lid and slits in the rubber sheets, gives the temperature of the water layer from which the sample is taken with an accuracy hitherto never obtained in hydrographic research.[17]

The thermometers did not have to make the journey through the sea and could therefore be delicate, accurate laboratory-type instruments. Pettersson's thermometer had scales divided to $\frac{1}{20}$ °C, allowing estimation to one-hundredth parts of a degree, vastly superior to the imprecise Miller–Casella thermometers. He was confident that the bottle could be brought up

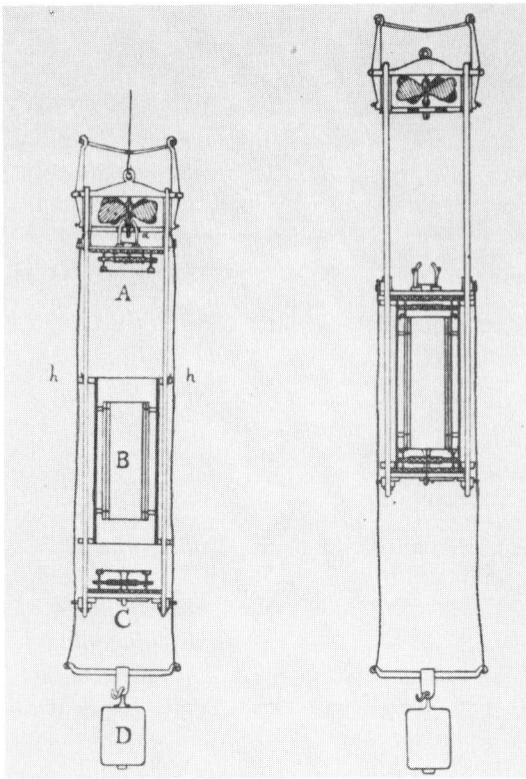

Figure 128. Pettersson's insulated waterbottle. A, upper lid; B, cylinder body; C, lower lid; hh, wires by which the weight helps to pull down the upper lid; D, weight. *Left*, the bottle descending, open; *right*, the bottle closed. From Pettersson 1894 *Scot. Geog. Mag.* **10,** plate 1.

from 600 m, even in unfavourable circumstances such as great variations in water or air temperature, with the temperature of the contents remaining constant for 1½–2 minutes.

The principle of insulation by concentric water layers had been adopted by Wauchope, although his primitive bottle also relied heavily on claddings (see §8.1). Pettersson took great care that his bottle should have the least possible capacity for heat, so that it quickly took up the ambient temperature, and he achieved this by substituting thin sheets of brass and rubber in place of the thick coats of tallow, canvas or gutta-percha previously favoured. With a small heat capacity, the bottle could be speedily lowered and raised, whereas the earlier bottles had had to be left hanging in order to come to thermal equilibrium with their surroundings.

Nansen took a bottle of this type on the *Fram* voyage and found its accuracy and speed of operation excellent. Observations were made through holes cut in the ice and the apparatus could be conveniently operated by one man—no small advantage on *Fram* whose small crew were kept busy with other scientific work. The expedition also took an Ekman bottle as a reserve instrument but Nansen found it heavy and awkward to use on a hemp line, uncertain in its closing and not reliable because the sample was sometimes diluted by fresh surface water.[18] After August 1894 he no longer used it, but when the following April Pettersson's bottle was lost after the sounding line broke, two of the crew made up a non-insulated bottle like Pettersson's in general form but closed by large rubber plugs. Nansen found it to be quite watertight, although its propeller release mechanism sometimes closed prematurely.

10.4. SPECIALISED DEVELOPMENT OF INSULATED WATERBOTTLES

After *Fram*'s return Nansen defined his requirements for a waterbottle, and suitable tests to prove its capabilities.[19] The extracts quoted below show how scrupulous was his attention to detail in his search for precision and reliability.

> It has generally been considered a satisfactory test, to let a waterbottle hang filled for some time, and to see whether any water ran out, if not, the bottle was declared to be tight.
>
> This, in my opinion, is no satisfactory test, for even if a bottle seemed tight when hanging thus, it is quite possible that it will allow a slow diffusion between the water inside and the water outside the bottle, while it is being hauled more or less rapidly through the water-strata.

A good test of the tightness of a water-bottle, in my opinion, is to fill it with distilled water, immerse it in the sea, and there move it about or haul it up and down, say for an hour or two. It must of course not be lowered so deep that the water contained is compressed by the overlying water. When the water-bottle is again taken up, it should not be possible to demonstrate the presence of salt in the sample of distilled water by adding drops of a solution of silver. I have made the experiment and found that it is most difficult to get a waterbottle so tight that it will stand this test quite perfectly, even if the bottle, hanging while filled, in the air for hours did not give off a single drop of water, and I feel convinced that in many cases, especially in previous expeditions, the waterbottles have been far from fulfilling these conditions.

Before the samples were taken, the waterbottle was as a rule wiped carefully dry outside near the stopcock, and in order to get rid of any drop of water that might be hanging in the mouth of the stopcock, this was opened, and a little water allowed to run out, before the water sample was taken in the glass bottle†.

Pettersson's bottle passed this severe test but Nansen had trouble with the detaching hook which sometimes froze solid, and with the lid-retaining hooks which became worn. In lesser depths, where temperature and salinity varied within short distances, he adopted a form of messenger release but this was time-consuming in deeper water and he therefore changed the propeller release, adding a shackle which could not slip off the hook prematurely.[20] He also added a pair of automatic locks to hold the lids and cylinder together when the bottle closed, to prevent the bottle springing open slightly when pounded by heavy seas.

At the same time Nansen corrected what he felt to be the bottle's main source of error.[21] Normally a thermometer was inserted when the bottle was recovered, and for this purpose a vent was provided through the insulation of the lid. Not only was the inserted thermometer probably at a different temperature to the water sample, but its insertion agitated the sample, thus speeding up heat transfer and reducing the efficiency of the insulation. Nansen therefore designed a long thermometer which could be clamped by washers in the lid and sent down with the bottle.

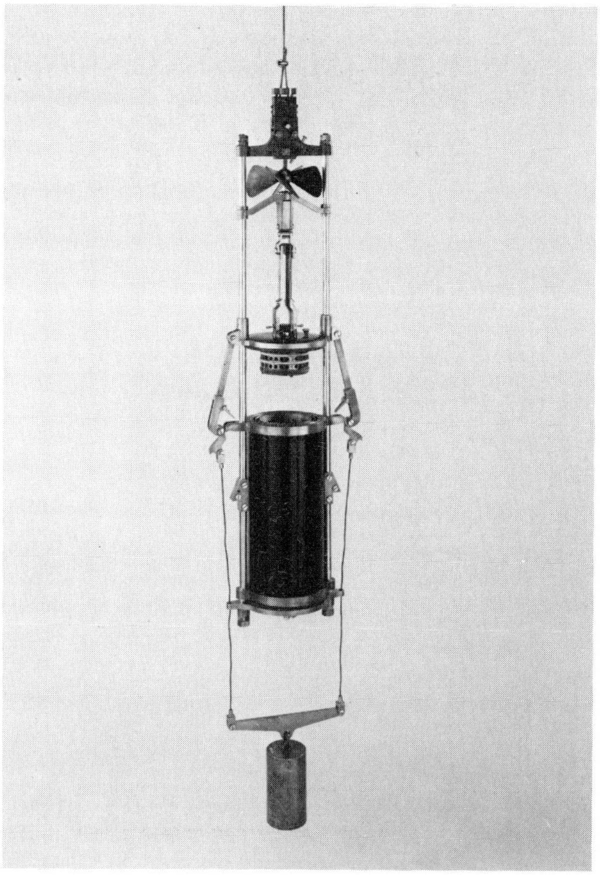

Figure 129. Pettersson–Nansen insulated waterbottle.

When the lid closed, the bulb lay within the sample and the scale could be read without disturbing the contents. This so-called Nansen deep-sea thermometer was first used on the *Michael Sars* under Nansen's supervision in 1900.[22]

The Pettersson–Nansen insulated waterbottle (figure 129) was used on several scientific expeditions around the turn of the century and was adopted by the ICES for distribution to its members. As constructed by the instrument-maker Ericsson of Stockholm, five concentric, insulating cylinders of brass and celluloid were fitted, separated by ebonite spacers.[23]

One of Nansen's first tasks as director of the Central Laboratory was to look into the insulating qualities of this bottle. His first report summarised the problem thus:

In the autumn of 1902 an examination of the properties of the Pettersson–Nansen bottle was made. An attempt to find the principles according to which a waterbottle of a

† This precaution was taken to avoid contamination by near-fresh ice meltwater which lay on the surface of the Arctic Ocean during the summer months. One or two drops in a sample of 400 cc reduced the specific gravity by 0.00001, enough to vitiate the observation.

certain size must be constructed to give the best possible insulating power, furnished data which will be the subject of a more detailed communication.[24] It was found that the insulating power could be increased by combining water jackets and solid walls of good insulating material. This arrangement would however have other inconvenient consequences—especially with the change of temperature of the solid walls caused by the great decrease of pressure when the bottle is hauled up, and so cause an error in the temperature determination which cannot be controlled. The old principle of insulating is therefore adhered to; however, the avoidance of direct metallic connection between the outer and inner parts has necessitated several alterations in the construction of the waterbottles.[25]

The sample was insulated to 0.01 °C for 8 or 9 minutes after which its temperature changed rapidly, but this satisfied Nansen's demands. Because of the uncontrollable thermal effect at depth, reversing thermometers were preferred to the larger Pettersson–Nansen bottle.

According to a later report of the Laboratory's test on the thermal characteristics of the bottle, the materials that went into its construction were 160 cm^3 brass, 180 cm^3 ebonite, 70 cm^3 celluloid and 130 cm^3 india-rubber. Added to this, and also vulnerable to adiabatic heating and cooling were the thermometer, 17 cm^3, and the contents, 1650 cm^3 of water. Calculation showed that when raised from 1000 m the adiabatic decrease in temperature would be 0.39 °C, largely due to the compressibility of the rubber, ebonite and celluloid components. The actual figure depended on the temperature of the water sample, being proportionately greater in warmer water. Such a deviation was totally unacceptable by Nansen's standards and two alternatives were considered: either the bottle should be reduced in size or it should be reserved for shallow-water observations, as Nansen had earlier recommended.[26]

The Laboratory backed the recommendation to discontinue the use of the larger bottle as its accuracy was not proportional to the higher insulating qualities and consequent increased weight. The smaller bottle might be improved by reducing the volumes of rubber, ebonite and celluloid in its construction. It continued to be popular and was adequate as long as certain precautions were taken; since it was used mainly in upper water levels it was normally triggered by a messenger.[27] On board *Michael Sars* a frame was fitted to the bottle so that a reversing thermometer could be attached, and when sent well below 1000 m, the reversing thermometer thus provided a check on the reading taken from the bottle.

His experiences with the Pettersson bottle led Nansen to design one of his own without any compressible materials, that would serve to collect deep samples without adiabatic cooling as it was hauled up. In his 'great Nansen insulated waterbottle' (figure 130) there were seven inner water-jackets, surrounded by evacuated glass tubes, each silvered internally. Compressible material was restricted to the small volume of rubber forming the lid plates.[28] As constructed by the instrument-maker L Andersen of Christiania, it had a capacity of 5–6 litres, and the inner

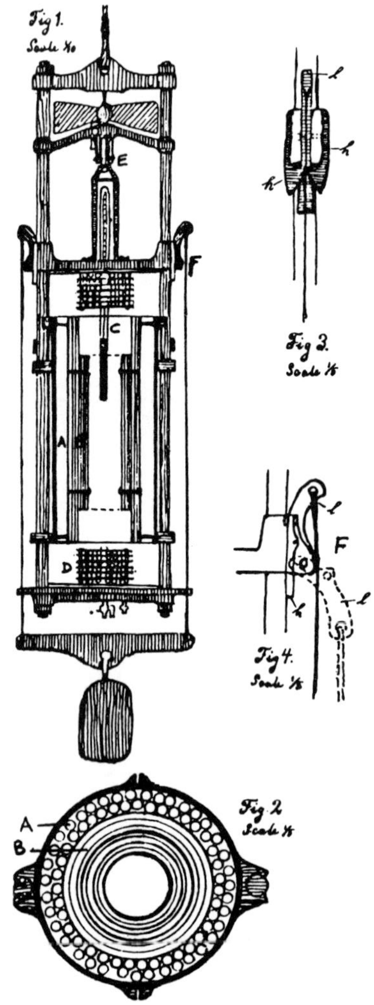

Figure 130. Great Nansen insulated waterbottle. A, outer ring of glass tubes; B, inner ring of brass rings; C, Nansen deep-sea thermometer; D, lower lid formed of separated plates of rubber; E, detaching hook; F, locking catch. From Nansen 1901 *Nyt. Mag. Naturv.* **39** plate 2, Figs 1–4.

and outer layers of water drained through separate taps. Nansen was pleased with its performance; his deep-sea thermometer fitted tightly into the lid, its insulation was excellent and the lever system attached to a suspended weight made it absolutely watertight. He found that when hauled up from 3000 m at a speed of 120–150 m per minute it retained temperatures to 0.01 °C, which implied that it insulated the sample for up to 25 minutes—a noteworthy performance. However, this bottle was lost at sea in August 1900 and Nansen did not have any others made to the same pattern.

A bottle similar in general design but insulated by an outer rubber jacket was designed by Nansen and Pettersson, and constructed by Ericsson of Stockholm.[29] It had no arrangement of eccentric levers to press down the lid and was therefore less watertight than the great Nansen waterbottle, but it was tested on *Michael Sars* and was reported to insulate samples from 3000 m 'with perfect accuracy'. No explanation was given as to the means of avoiding the adiabatic cooling which would be expected from soft or hard rubber at this depth.

To increase the number of observations that could be made in a limited time the Central Laboratory also developed a bottle for use on ships under way. Such a bottle would be in motion all the time it was in the water and, streaming out behind the ship, would require some means of registering the depth at which it closed, since the length of tow-line paid out would greatly exceed the vertical depth at which the bottle was running.

The bottle was designed by Nansen and V W Ekman, the youngest son of F L Ekman, inventor of the insulated waterbottle described in §10.3. Known as the Nansen–Ekman or 'automatic' bottle, its design owed more to the series of non-insulated bottles put out by the Laboratory at this time (figure 131). Its slender cylinder was clad by 2 mm of ebonite and contained two inner water-jackets. There was no frame, the tow-line being attached to a yoke, and the lids, insulated with inner rubber pads and outer ebonite covers, were linked to a hinged flap counterpoised by a suspended weight. An ordinary thermometer was carried within the bottle and the closing depth was given by a Rung bathometer (a glass tube pressure gauge) fastened to the cylinder in such a way that its tap was shut by the bottle's closing mechanism. It showed the level from which the sample was taken to an accuracy of 1 m.

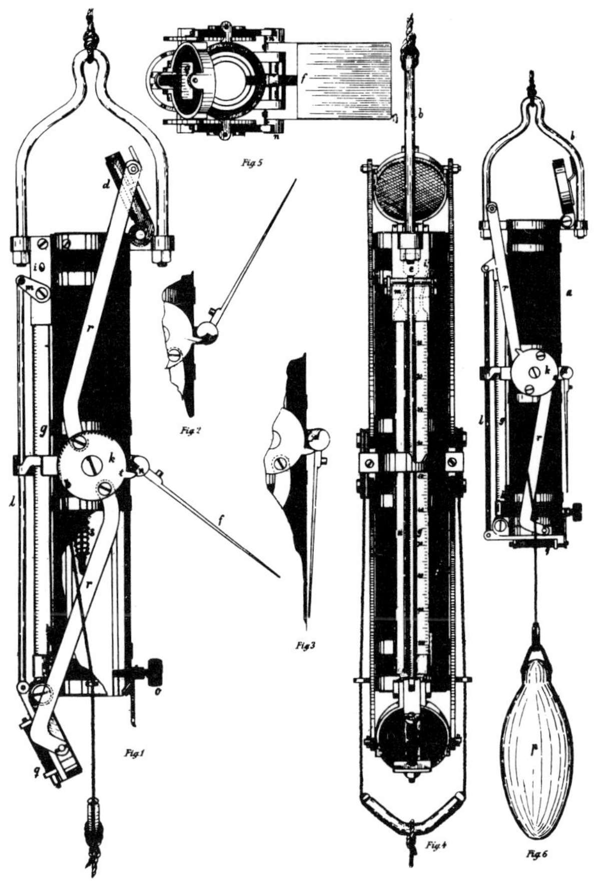

Figure 131. 'Automatic' insulated waterbottle. From Ekman 1905 *Cons. Perm. Int. Expl. Mer Publ. Circ. No.* 23, plate 2.

Unlike Pettersson's bottle, the tap released water from the jackets as well as from the inner cylinder, but as the outer water emerged first it served to clear the tap of any contaminating surface water and to wash out the intended storage container. It was found to insulate for two to three minutes, and operated well down to 60 m from a vessel steaming at seven knots. It had been designed and tested during 1903–4 and was recommended for use the following year.[30] Subsequently a larger model with a capacity of 800 cc was produced. Four concentric jackets were provided and the insulation held for four to five minutes.

Nansen's experiences with this bottle on board *Frithjof* were satisfactory, but below 200–300 m the ship had to heave-to as the winch could not take the quantity of line needed to stream the instrument at its maximum operating depth of 600 m. With such a bottle large areas of sea could be explored at regular intervals throughout the year without excessive demands on the time of ships and crews. As the automatic bottle was limited in its range, Martin Knudsen took the design further and in 1909 published details of his 'high-speed' bottle (figure 132).[31] Knudsen retained the slender form of the automatic bottle but, finding that at high speeds water rotated within the jackets and destroyed the insulation, he was forced to insert cross-pieces even though this led to heat exchange by conduction. The lids were linked by rods and springs with the closing action triggered only by the strong motion of hauling up and in, added to the forward speed of the ship. Two pressure tube bathometers were fitted, one calibrated for shallow depths and one for deep water. The insulation held for two minutes so that the action of hauling in and reading the thermometer had to be very rapid.

Knudsen's trials were made during a crossing from Stavanger to Aberdeen. With the ship travelling at 8 knots he was able to sound to the bottom, an average depth of 80 m, all the way across, except in the deep trench adjacent to the Norwegian coast. Here, between 180 and 250 m, he got casts by slowing the ship to 2 knots. In all, 470 observations of temperature and salinity were taken in the course of the 37½ hour crossing.

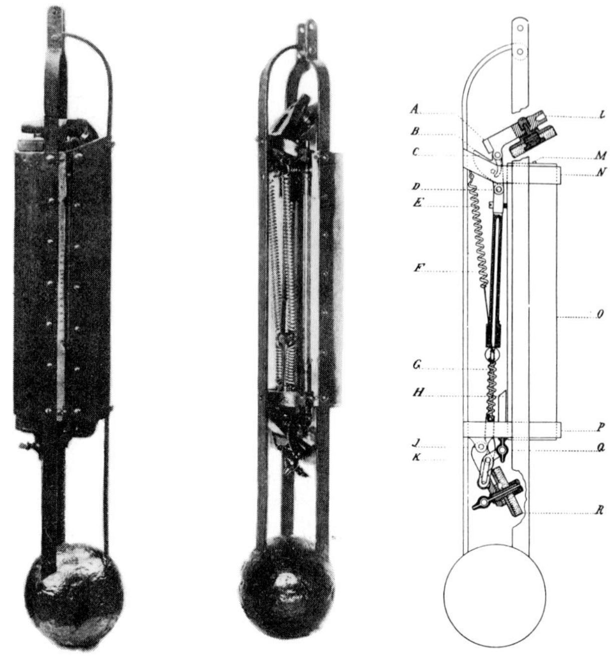

Figure 132. Knudsen's high-speed bottle for use by ships under way. *Left*, the bottle in its casing ready for use; *centre*, with the casing removed to show the springs which operate the end-valves; *right*, section drawing. From Knudsen 1909 *Cons. Perm. Int. Expl. Mer Publ. Circ. No.* 50.

10.5. NON-INSULATED WATERBOTTLES OF EKMAN (1869–76) AND NANSEN (1900–2)

Ekman's deep-water bottle No. 1 was made for Baltic Sea investigations in 1869 (figure 133).[32] To prevent any contaminating air or water being carried down from upper levels, Ekman's apparatus consisted of a cylinder borne up by water thrust above its closing plates, onto which it dropped when it reached the sampling level. The original model was only 40 cm long, its frame was of tinned iron and the cylinder, thrust plates and end plates of brass. The cylinder dropped into tallow-filled grooves in these endplates, which then acted as a watertight seal. The design was less than ideal, however, since the bottle tended to close prematurely and, on one occasion, having gone down closed but empty, had been crushed by pressure. Ekman therefore added a hook to support the cylinder until it was moving freely down through the water.

Ekman's deep-water bottle No. 2, made to hold 1.3 litres, and with a slightly altered frame and base plate, was exhibited at the Geographical Congress in Paris in 1875 where it was commended by the jury (figure 134).[33] The same model was shown at the Philadelphia exhibition of 1876, at the Fisheries exhibition in Berlin in 1880 and at the Geographical Congress in Venice in 1881.[34] In practice, however, Ekman found it difficult to balance the weight of the cylinder against the speed of descent, so that it did not close prematurely.

Several modifications were introduced in his bottle No. 3. This was a sturdy apparatus; three steel guide rods were provided and three spring-steel feet protected it from damage at the sea bed. The greased seals were replaced by a rubber pad and washer, and a hook below the base plate caught the rim as it closed, for greater

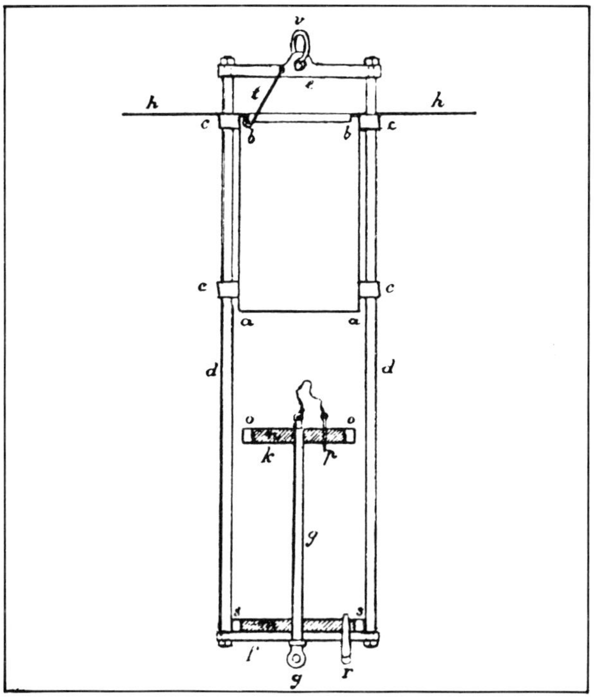

Figure 133. Ekman's No. 1 bottle made for the 1869 Baltic expedition. Redrawn from Ekman and Pettersson 1893 *Kong. Svenska Vet. Akad. Hand.* **25** No. 1 Fig. 1.

security on the way up. The brass cylinder was tapered, so that the water forced it against the seals as it was raised.

Ekman thought it a good design and to make it reliable even in 'oceanic' depths he fitted a propeller and ratchet mechanism which only allowed the cylinder to fall after the bottle had been hauled upwards for at least 20 m, a distance greater than any wave turbulence. This bottle No. 4 (figure 135) was also shown at Berlin and Venice. Ekman suggested that the propeller could also be made to register the depth travelled, which would be useful in strong currents, but he was deterred from making such a meter because of the high cost involved.

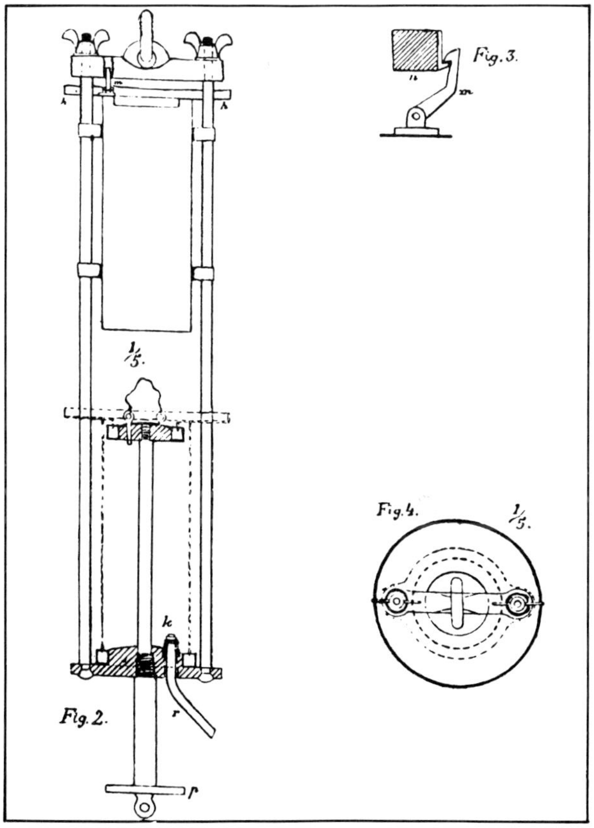

Figure 134. Ekman's deep-water bottle No. 2. From Ekman and Pettersson (1893) Fig. 2.

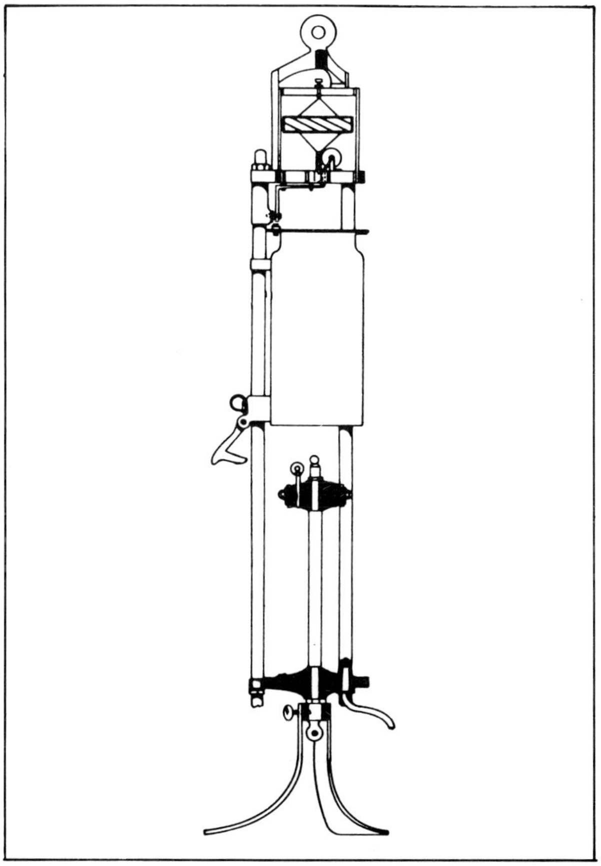

Figure 135. Ekman's No. 4 bottle fitted with propeller-operated release mechanism. Redrawn from Ekman and Pettersson (1893).

In deep water, where insulating bottles were still impractical because they took too long to haul up, reversing thermometers were used. Water samples for analysis were still needed, but these could be brought up in small, simple bottles. These were attached individually to the line or paired with the thermometers in a reversing frame, and Nansen had several such bottles made for his cruise in 1900.[35]

Figure 136 shows two stopcock bottles hinged to a rod and released by a propeller mechanism, the dropping action closing all stopcocks simultaneously. The bottle cylinders had conical ends to ease the free flow of water as they descended, thus yielding a sample truly representative of the closing level. Nansen experimented on the same voyage with a barrel-shaped bottle which could be arranged on the line opposite a reversing thermometer frame. It held more than half a litre and Nansen considered that it too allowed the water to pass freely through during its passage down to the sampling level. On the *Frithjof* cruise of 1910 Nansen took two other instruments consisting of a stopcock bottle and a reversing thermometer assembly.[36] He reported that they worked perfectly well but gave no description of their construction.

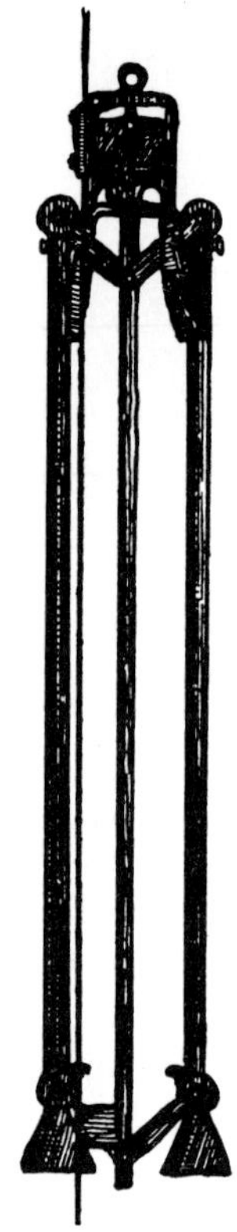

Figure 136. Nansen's stopcock waterbottle. From Nansen (1901) plate 2, Fig. 5.

10.6. PRINCE ALBERT OF MONACO AND HIS APPARATUS

His Highness Prince Albert I of Monaco (1848–1922) must surely have been the only crowned head to make marine research his lifelong pursuit. He began his career in 1873 with the acquisition of the schooner *Hirondelle* and in 1885 embarked on a serious study of oceanography in all its aspects. In 1889 he became head of the Principality of Monaco, a territory almost reduced to that occupied at the present day, and one whose population included many fishermen and others making a living from the sea.

Prince Albert's interest in zoology led him to study marine fauna against the background of the physics and chemistry of sea water. He broke much new ground, introduced new apparatus into this work, and towards the close of his life, in a lecture to the National Academy of Sciences in Washington, listed among his achievements various closing nets and photometers to measure the penetration of light into water. He worked systematically over his chosen areas, returning each season until satisfied that he had learned as much as possible about conditions there. His crews' ability to operate nets at a known depth was extended to the similar operation of waterbottles, thermometers and sounding gear which backed up his biological investigations. He not only commanded his various ships but directed the oceanographic work, inviting noted specialists to join him. One of these was J Y Buchanan, who had served as chemist on board HMS *Challenger*.

In addition to the conventional publication of his results, Prince Albert made two major contributions to the young science of oceanography, the first of which was the foundation of the Oceanographic Museum in Monaco. Dramatically situated on a cliff at the sea's edge, the museum building was begun in 1899 and formally opened in 1910. Working laboratories occupied its lower floors, above which the Prince set out exhibition galleries to display his own apparatus alongside that presented by other governments and individuals. A vast number of sea creatures were preserved for display, as well as manufactured items derived from marine products. The Prince later founded an Oceanographic Institute in Paris and courses under Professor Thoulet and others were held there. Both Museum and Institute founded regular journals to publicise their work.

The second undertaking which owes its origins and its present base in Monaco to Prince Albert's initiative is the International Hydrographic Bureau. Many national bathymetric charts were produced during the late nineteenth century based on data from telegraph soundings and scientific cruises. In 1902 the Seventh International Geographical Congress, meeting in Berlin, resolved to compile a world bathymetric chart and Prince Albert undertook to have it drawn and printed in an edition of 24 sheets at a scale of 1:10 million. The finished work was ready in time for the Eighth Geographical Congress held in New York in 1904, and a second edition was prepared during 1910–12. In 1921 the International Hydrographic Bureau was set up to coordinate the activities of all national hydrographic offices.

In 1885 the Prince made his first cruise in the Gulf of Gascony, that portion of the Bay of Biscay which lies near the Franco–Spanish border, and in succeeding years his voyages extended out towards the Azores. Prince Albert made a study of the Gulf Stream in the eastern Atlantic and, in order to see if its surface drift reached the shores of Europe, he released each year a number of marked floats of his own design, together with various barrels and bottles, which, if recovered and returned, would yield this information. He came to rely on copper-clad glass spheres, loaded with lead shot so that they barely floated, and containing messages in nine languages, soliciting their return with a note of the date and place of their recovery (figure 137).

The *Hirondelle* was entirely without steam and all her winding gear was hand-operated. For sounding, a Baillie detacher, modified by Jules Leblanc, was used from 1888 onwards. A tap or key replaced the butterfly valves normally fitted to the Baillie tube, and gave the sounder its name *sondeur à clef* (figure 138).

The four seasons spent aboard *Hirondelle* were so rewarding that Prince Albert commissioned from an English shipbuilder the larger and better-equipped 600-ton steam-yacht which he named *Princesse Alice*. This took him in 1892–93 through the western

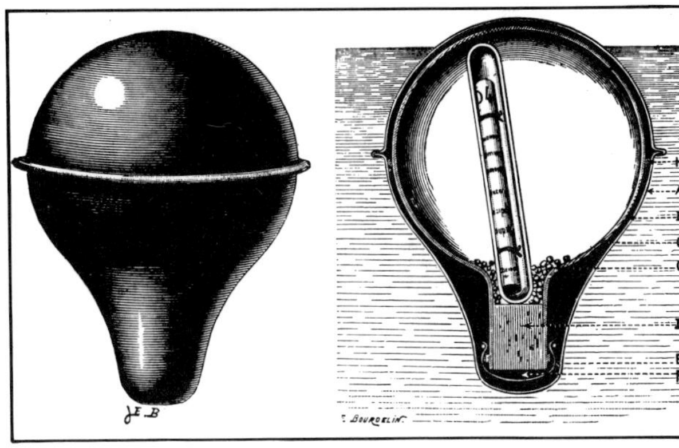

Figure 137. Copper-clad glass spheres. From Richard 1900 *Les Compagnes Scientifiques de SAS le Prince Albert . . .* Figs 13 and 14.

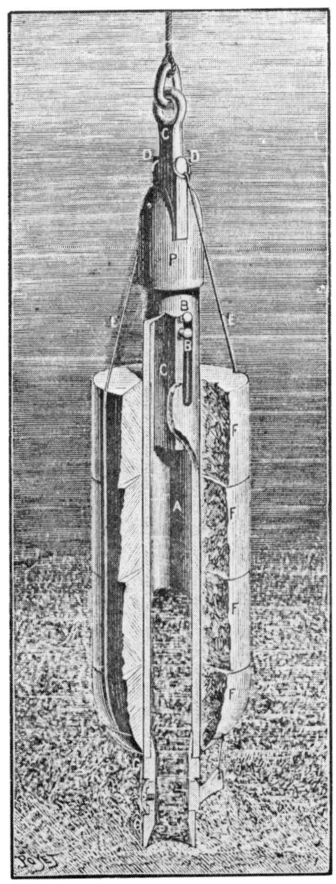

Figure 138. Travailleur sounder (*sondeur à clef*). From Richard 1900 *Les Compagnes Scientifiques de SAS le Prince Albert . . .* Fig. 4.

Mediterranean. Between 1894 and 1897 he returned both to the Mediterranean and to the Azores.

In 1892 Buchanan, assisting the Prince, substituted a messenger in place of the flap closure on his stopcock waterbottle. He may have been encouraged by the French preference for messenger-operated equipment, for in 1881 a stopcock bottle triggered in this way had been made for *Travailleur*, a French zoological research ship. Unable to obtain Buchanan stopcock waterbottles in time for their 1881 season, Captain Richard and Lt Villegente designed and made the steel *Travailleur* bottle, six examples being used at sea that year.[37] Intended to collect mid- and bottom-water samples, one or more bottles were sent down simultaneously with the dredge, but on a separate line (figure 139).[38] The bottle went down open, each stopcock holding open the adjacent valve, and then a messenger weighing 15 or 20 kg was sent down the line. Passing over the bottle, it knocked each lever down in turn, closing the stopcocks and releasing the valves which clamped ever more securely as the bottle rose and its internal pressure exceeded that outside. A single messenger could pass over a sequence of bottles, triggering each in turn. The water flow through these bottles was hampered by the tight bends and narrow orifices within. Folin, the scientist on board, thought that the speed of descent overcame this defect,[38] but Richard considered that the water brought to the surface did not truly represent that of the closing depth.

It was often remarked that deep-water samples foamed when opened, due to gas bubbles forming under the sudden release of pressure. For example, Milne-Edwards, opening his bottles on *Travailleur*, saw on several occasions a metre-long jet of water expelled from the bottle, and the water emptied into a receiving flask was full of bubbles. It was some time before scientists recognised that this was not gas that had been dissolved in deep sea water but simply air that had been trapped within the bottle as it entered the sea. Buchanan avoided this nuisance by keeping his stopcock bottle shell but altering its closure. He described the new release mechanism to the Royal Society of Edinburgh the following winter (figure 140):

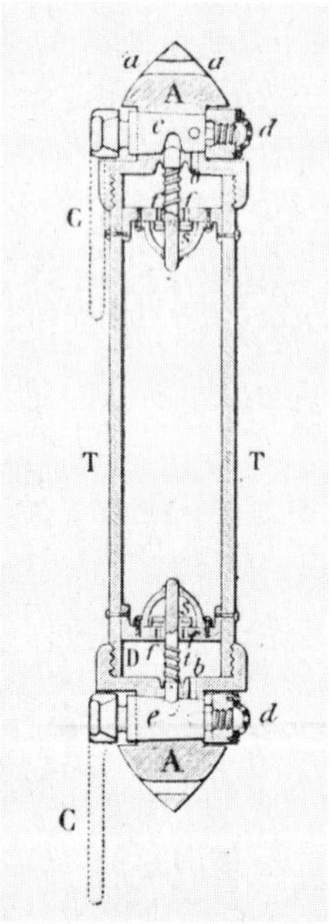

Figure 139. Travailleur waterbottle. From Milne–Edwards 1882 *La Nature* Pt 1 p53, Fig. 3.

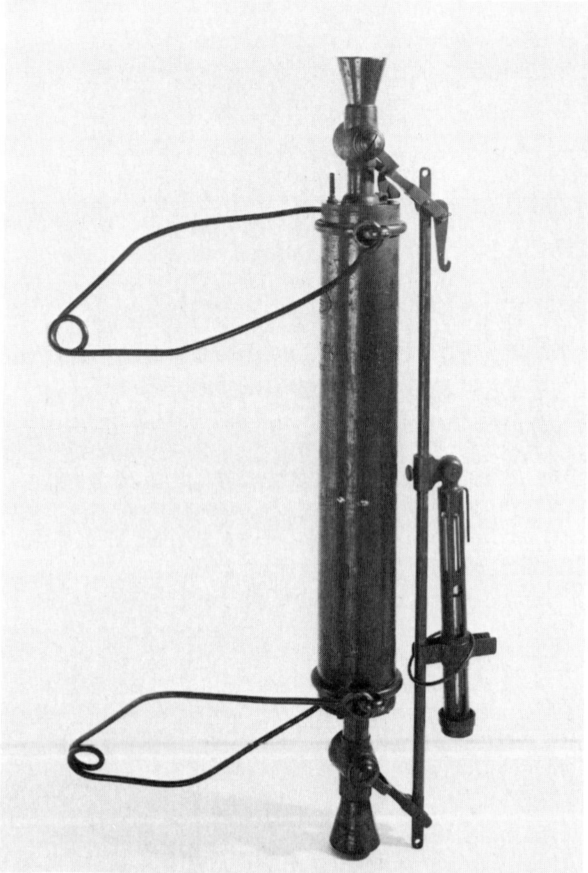

Figure 140. Buchanan stopcock waterbottle used on the 1901 Antarctic expedition. To the right of the cylinder is a rod carrying the reversing thermometer.

Where the messenger is used no skill and experience are required. [On *Challenger* he had found it necessary to supervise all waterbottle operations himself.] The stopcocks are supported by a line or wire from the upper end of the connecting-rod to the trigger. When the messenger falls on the trigger, the stopcocks close by the weight of the connecting-rod, and, if necessary, attached weights. . . . On board the *Princesse Alice* the messenger used was very heavy. It consisted of a ring of lead, weighing rather over two kilogrammes. Its impact on the trigger, at a depth of 1000 metres, was distinctly felt at the surface, wire being used.[39]

It took over 12 minutes for the messenger to slide down 1000 m of wire. On a line out of plumb the messenger might fail to reach the target, but Buchanan felt that this was an advantage since all observations made from a non-vertical line were probably best discarded.

When a thermometer and waterbottle were separately attached to the line it was a lengthy business to detach and read the thermometer, to empty the bottle and to re-attach them both at different positions. Buchanan arranged the thermometer on the bottle frame, so that it was capsized as the bottle closed, and then clamped in the reverse position while hauled in. In this form the bottle served on several expeditions, notably the Royal Navy's Antarctic expedition of 1901–4, and the *Scotia* expedition of 1902–4, the Scottish expedition led by W S Bruce. They were used with 'Knudsen improved' pattern thermometers, made by Negretti and Zambra, which by then had been discarded by the Scandinavian workers as unreliable and inaccurate.

The *Princesse Alice*, despite her modest size, held laboratories and all the gear necessary for efficient work, including a sounding machine designed by Le Blanc. It was of conventional pattern but vertical configuration, dictated by its site on deck, and it could be powered by steam or electricity. Work in the eastern Mediterranean was directed largely to measurement of temperature and salinity, and therefore involved numerous sounding operations. In 1894 Prince Albert, dissatisfied with the brittle quality of his piano-wire, replaced this by a three-strand cable, 2.3 mm diameter, made up of nine galvanised steel wires. The greater reliability of such a cable (its breaking strain was around 250 kg) allowed him to put series of bottles and other apparatus on the line with confidence.

Feeling that he had learned as much as could be expected from the Azores region, the Prince took his new ship *Princesse Alice II* polewards in 1898. Powered by a 1000 hp engine, and carrying 245 tonnes of coal, the vessel could steam at 13 knots. She was rigged as a two-master and was 73 m long. Fore and aft steam and electric winches were provided. On this voyage from Norway to Spitzbergen and back via the Faeroes, Prince Albert collected larger animals from land and sea, besides making a study of plankton, and he returned to the Arctic the following year after a preliminary sortie into the Mediterranean.

Jules Richard, who became Director of the Oceanographic Museum in Monaco, developed much of the apparatus used by the Prince, and in 1902 he described a small stopcock bottle in a Magnaghi frame (figure 141).[40] The entire bottle could be released by propeller or messenger, to capsize about its axis. Several bottles could be attached to the line if they were propeller-operated, but there appears to have been no provision for releasing messengers from one to the next. Léger and Taffe of Monaco made the bottles, which were kept small as the stopcock had to be the same size as the cylinder if water was to pass freely through during descent.

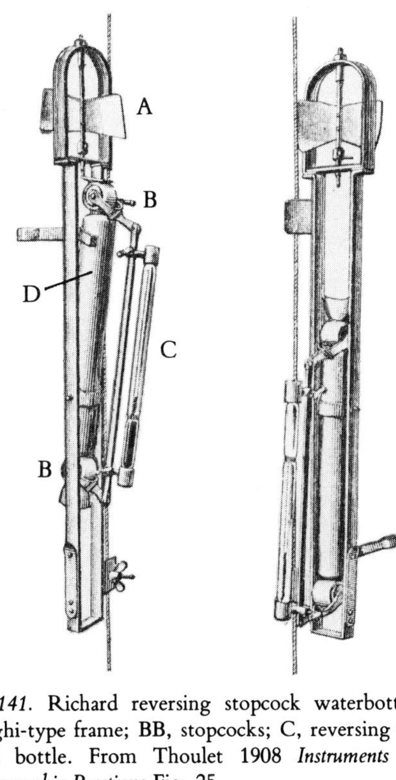

Figure 141. Richard reversing stopcock waterbottle, 1902. A, Magnaghi-type frame; BB, stopcocks; C, reversing thermometer; D, the bottle. From Thoulet 1908 *Instruments et Opérations d'Océanographie Practique* Fig. 25.

In 1908 Professor Thoulet of Nancy University praised this bottle in his textbook of practical oceanography and at the same time illustrated a simplified version which he recommended (figure 142); it was smaller in size but adequate for samples for gravity and chemical analyses.[41] The Thoulet bottle was in effect a normal reversing thermometer to which a light bottle could be added at will. It went down weight uppermost, with the valves open. As the two instruments capsized, the lead weight dragged down the central rod, moving the valve plates close to the cylinder and trapping a water sample. By allowing the weight to drop slightly, air then entered the cylinder and the water drained out. The crudeness of this method contrasts sharply with that of Nansen and his co-workers in northern seas at this time, although it may have been adequate for Thoulet's purposes; he criticised other available bottles, such as those of Mill, Buchanan and Pettersson, as being too costly.

For sediment sampling on board the ships of Prince Albert and contemporary French expeditions there were alternatives to the *sondeur à clef* which served mainly in deep water over a soft bottom. For example, in lesser depths and where a sea-bed water sample was required, the Buchanan sounding tube was sent down. This composite apparatus, designed by Buchanan after *Challenger*'s return, combined a central sampler tube with a jacket which captured a water sample as the detacher slipped off and the tube was lifted from the sea bed. No tube sounder functioned in gravelly deposits and here the Légér grab, a small jawed device, was normally used. In very deep water other types of dredge might be the only effective way of gathering material from the bottom.

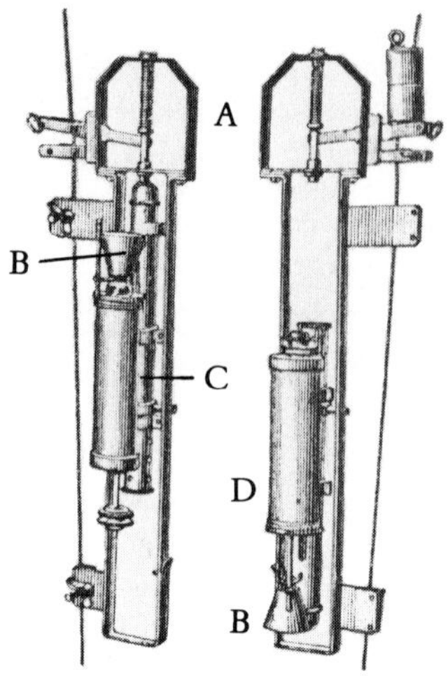

Figure 142. Thoulet reversing stopcock waterbottle, 1908. A, 'Scottish' frame; B, lead weight; C, reversing thermometer; D, the bottle. From Thoulet 1908 *Instruments et Opérations d'Océanographie Practique* Fig. 26.

10.7. ICES REVERSING BOTTLES

In 1902 the Central Laboratory constructed a non-insulated reversing waterbottle which resembled their automatic bottle in general form. According to Nansen:

> A waterbottle specially adapted for use with the reversing thermometers has been invented. The cylindrical bottle which carries the thermometer case is reversed by its own weight, and by an eccentric mechanism the india-rubber ends are pressed tight against the two extremities of the cylinder. The bottle has been found to work quite well. With a capacity of rather more than 500 cc it weighs 4 or 5 kg . . . It can be obtained adapted either for propeller or messenger.[42]

Development continued the following year:

> A new pattern of W Ekman's waterbottle . . . has been constructed. This is for fixing to the side of the line and has an arrangement for releasing a messenger below, on reversing. In this way any number of bottles may, as first proposed by Mill, be put on the same line.[43]

The bottle was made to two patterns: one to be clamped beside a line, with an arrangement for releasing another messenger below; the other to be fixed at the end of a line. Ekman described its construction and operation:

> The water sample is collected in the brass tube (tinned inside) pivoted on an axis. The lids, movable on hinges, are closed and opened by two pairs of eccentric rods. The waterbottle is sent down with the tube in its uppermost

position, where it is kept fastened by the small bolt; the ends are then open, and the water runs quite unhindered through the tube. When the messenger arrives, the tube is released, and given an impulse away from the vertical by the spring then by its own weight it falls right down to the lowest position where it is caught and fastened by the hook; the lids are then closed quite tightly.[44]

The capsizing action released the messenger which travelled to the bottle below and the operation was repeated as often as necessary. Holes in the lids allowed the water sample and rubber liners to expand slightly as the instrument was raised. Two teeth were provided on each lid so that even when the packing became slightly worn the lid could still be tightly clamped. Originally one, and later two reversing thermometers were attached to the bottle frame. Two sizes were made, by Ludvig Andersen of Christiania, the smaller holding 520 cc and the larger 1070 cc of water (figure 143).

During the same year, Norwegian Royal Navy observers on the Isachsen expedition to Spitzbergen took an old model Pettersson insulated bottle, a small Pettersson–Nansen, and an Ekman reversing water-bottle.[45] The reversing bottle was used in levels below 600 m and often at 400 m and 500 m also. It did not always close properly so that some of the results were useless as the sample was diluted. Furthermore, both the reversing thermometers intended for use with this bottle were unreliable, but the older insulated bottles were apparently satisfactory.

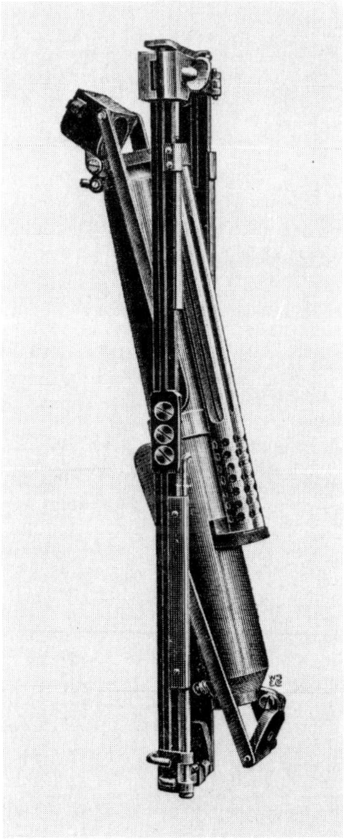

Figure 143. Ekman's reversing waterbottle offered for sale by Ludvig Andersen of Christiania, *c*1910. From Andersen's Catalogue, 1910.

REFERENCES AND NOTES

1. Pettersson O 1894 'Proposed scheme for an international hydrographic survey of the North Atlantic, the North Sea and the Baltic' *Scot. Geog. Mag.* **10** pp631–4
2. Pettersson O 1895 'Proposed scheme . . .' *Rep. 6th Int. Geog. Congress* (London) pp587–90
3. Dickson H N 1894 'Report on the physical investigations carried out on board HMS "Jackal" 1893–4' *12th Ann. Rep. Fisheries Board for Scotland* pp336–82
4. Nansen F 1893 'On how the polar ocean might be crossed' *Geog. J.* **1** pp1–32
5. *The Norwegian North Polar Expedition* 1901 (6 vols). See especially Nansen F 'The oceanography of the North Polar Basin: scientific results' in vol. 3 No. 9 pp1–247
6. Denmark: *Report of the Danish 'Ingolf' expedition* 1899 (Copenhagen)
7. Grützmacher F 1904 'Über Tiefsee—Umkippthermometer' *Zeitschrift für Instrumentenkunde* **24** pp263–8
8. Norway: *Report of the Norwegian Fishery and Marine Investigations* 1900 vol. 1, Introduction, pp1–12
9. Norway 1900 *Op. cit.* vol. 1 No. 2
10. Went A 1972 'Seventy years a-growing: A history of ICES' *Cons. Perm. Int. Exp. Mer.* **165**
11. Meyer H A 1871 *Untersuchen über Physikalischer Verhältnisse des Westlichen Theiles der Ostsee* (Kiel) pp12–14
12. Ekman F L and Pettersson O 1893 'Om den Svenska hydrografiska expeditionens af ar 1877 . . .' *Kong. Svenska Vet. Akad. Hand.* **25** No. 1
13. Ekman F L 1881 *Appareils exposés par le Professeur F L Ekman au congrès géographique de Venise 1881* (Stockholm)
14. Philadelphia, International Exhibition 1876 *Swedish Catalogue* vol. 1 Statistics pp65–6
15. Pettersson O 1894 'A review of Swedish hydrographical research in the Baltic and North Seas' *Scot. Geog. Mag.* **10** p283
16. Richard J 1907 *L'Océanographie* (Paris) p107
17. Pettersson O 1894 *Op. cit.* p284

18. Nansen F 1901 *The Norwegian North Polar Expedition* vol. 3 No. 9 p141
19. Nansen F 1901 *Op. cit.* p136
20. Nansen F 1901 *Op. cit.* pp138–9
21. Helland-Hansen J and Nansen F 1909 *The Norwegian Sea: Its Physical Oceanography* (Christiania) p45
22. Helland-Hansen J and Nansen F 1909 *Op. cit.* pp48–50
23. Ekman V W 1905 'On the use of insulated thermometers and waterbottles' *Cons. Perm. Int. Exp. Mer Publ. Circ. No.* 23 p14.
24. This appeared as *Publ. Circ. No.* 23, 1905 (see ref. 23)
25. Nansen F 1903 'Report of the Central Laboratory for the year 22 July 1902 to 21 July 1903' *Cons. Perm. Int. Exp. Mer Rapports et Procès-verbaux* **1** Annexe A, ppXL–XLIV
26. Ekman V W 1905 *Op. cit.* pp13–21
27. Murray J and Hjørt J 1912 *The Depths of the Ocean* (London) pp232–3
28. Helland-Hansen J and Nansen F 1909 *Op. cit.* p51; and Nansen F 1901 'Some oceanographical results of the expedition with the "Michael Sars" in the summer of 1900' *Nyt. Magazin for Naturvidenskaberne* **39** pp101–3
29. Nansen F 1901 *Op. cit.* p133 (see ref. 28)
30. Nansen F 1905 'Report of the Central Laboratory for the year 22 July 1904 to 21 July 1905' *Cons. Perm. Int. Exp. Mer Rapports et Procès-verbaux* **4** Annexe A, pXXIII
31. Knudsen M 1909 'Ein Wasserschopfer zur Benutzung wahrend der Fahrt des Schiffes' *Cons. Perm. Int. Exp. Mer Publ. Circ. No.* 50
32. Ekman F L 1870 'Om salthalten i Hafsvattnet utmed Bohuslanska kusten' *Kong. Svenska Vet. Akad. Hand.* **9** No. 4, 44pp
33. Paris: *Congrès International Géographique* 1875 vol. 2 p414 (Paris). See also ref. (43)
34. Ekman F L 1876 'Description of hydrographical and meteorological instruments exhibited by Götheborgs och Böhuslans Hushallningssällskap at the Philadelphia Exhibition 1876, invented by F L Ekman'; and Berlin: Fischerei-Ausstellung Wissenschaftliche Abtheilung *Instrumente zur Untersuchung des Wassers* pp1–54, and Ekman F L *Op. cit.* 1881 pp4–5
35. Helland-Hansen J and Nansen F 1909 *Op. cit.* p55
36. Nansen F 1913 *The Waters of the Northeastern North Atlantic. Investigations made during the Cruise of the 'Frithjof', of the Norwegian Royal Navy in July 1910* (Christiania) p7
37. Milne-Edwards H 1882 'La campagne de dragages du "Travailleur"' *La Nature* pp52–3
38. Folin L de 1887 *Sous les Mers: Campagnes d'Exploration du 'Travailleur' et du 'Talisman'* (Paris) p12
39. Buchanan J Y 1893 'On some modification of the waterbottle and thermometer for deep-sea research' *Proc. R. Soc. Edin.* **19** p239
40. Richard J 1902 'Sur une nouvelle bouteille destinée a recueillir l'eau de mer à des profondeurs quelconques' *C. R. Acad. Sci. Paris* **134** pp1385–7
41. Thoulet J 1908 *Instruments et Opérations d'Océanographie Practique* (Paris) pp82–3
42. Nansen F 1903 *Op. cit.* pXLIII
43. Nansen F 1904 'Report of the Central Laboratory for the year 22 July 1903 to 21 July 1904' *Cons. Perm. Int. Exp. Mer Rapports et Procès-verbaux* **2** Annexe A, pXXI
44. Ekman V W 1905 *Op. cit.* pp27-8
45. Helland-Hansen J and Nansen F 1912 *The Sea West of Spitzbergen* (Christiania) p113

Chapter 11

Last on the Scene: The Deep-water Current Meter

11.1. MEASURING THE DIRECTION AND VELOCITY OF CURRENTS

Only passing reference has been made previously to the measurement of currents because the current meter, in the strict sense, did not come into use until the closing years of the nineteenth century. As late as 1954 the German hydrographer Bohnecke declared:

> The subject of current measurements has kept the oceanographers busy for more than a hundred years without having found—this must honestly be admitted—an entirely satisfactory solution.[1]

A current has both direction and velocity; these parameters may be measured at a single instant, over a measured period, or for an unspecified time. If a current is to be measured from a ship out of sight of land, then the observed result must be corrected by taking into account the ship's movement, whether this be due to wind or surface current. Since anchoring in deep water is seldom feasible, it will be seen that obtaining precise measurement even of surface currents is extremely difficult, while by measuring a subsurface current the ship will alter her own way through the water, rendering such corrections more problematical.

In shallow seas and in the upper levels of the ocean currents flow under the mean influence of wind and of tides. In lower levels they are set up by local differences in salinity and/or temperature. In all cases they are subject to topographic control and their flow may be concentrated and accelerated by the shape of nearby coastline or the sea bed, and in many instances they are also deflected by the Coriolis force. It will be apparent that surface currents are no guide to those beneath them; surface currents may vary from hour to hour and seasonally, in direction and velocity, while deep currents are more steady in character. The result of any observation therefore depends in part upon the time period over which it is made.

The current meter may register direction, velocity, or both parameters. The simplest form is a passive drifter which will subsequently be recovered. Recording meters may be divided into those which register within the apparatus itself and those which telemeter information back to the surface.

11.2. DRIFT BOTTLES AND FLOATS

In the chapter concerning Arctic voyages the instructions to commanders, ordering the daily release of marked floats were referred to. During the first half of the nineteenth century such experiments were numerous, and summaries of those recovered from the North Atlantic were published by Commander Becher in 1843 and 1857, preceded by many brief notices in the *Nautical Magazine* (figure 144).[2] Another chart gave the positions of icebergs and ice fields seen by British and American navigators in the Atlantic between 1832 and 1844.[3] Over 100 of these positions were marked and the fact that some of them lay within the supposed limits of the Gulf Stream led scientists to postulate a subsurface counter-current which impelled the submerged bulk of the iceberg.

In confined seas floats reached their destinations more quickly and gave a rapid indication of prevailing drifts. Aimé systematically released floats in the Mediterranean,[4] as did surveyors working along the eastern coast of North America around the middle of the century.[5] Between 1885 and 1887 Prince Albert of Monaco released 1675 barrels, bottles and specially made copper floats of various patterns, to discover if the Gulf Stream extended to the shores of Europe.[6]

Weighted drifters with a density greater than that of sea water can be released to drift along the sea bed. The United States Coast Survey (USCS) released such drifters made of cork and cement in 1885, but found that they recovered a higher proportion of marked seashells.[7] From this work, the USCS were able to identify residual currents over the inshore sea bed, as distinct from the daily ebb and flow of the tides.

Uncontrolled drifters show only average current direction but not its velocity. At the surface, they are at the mercy of wind, while those at the sea bed risk getting snagged on rocks or seaweed. By using floats

Figure 144. 'Bottle chart' of the Atlantic Ocean, 1843. From Becher 1857 *Naut. Mag.* unnumbered plate.

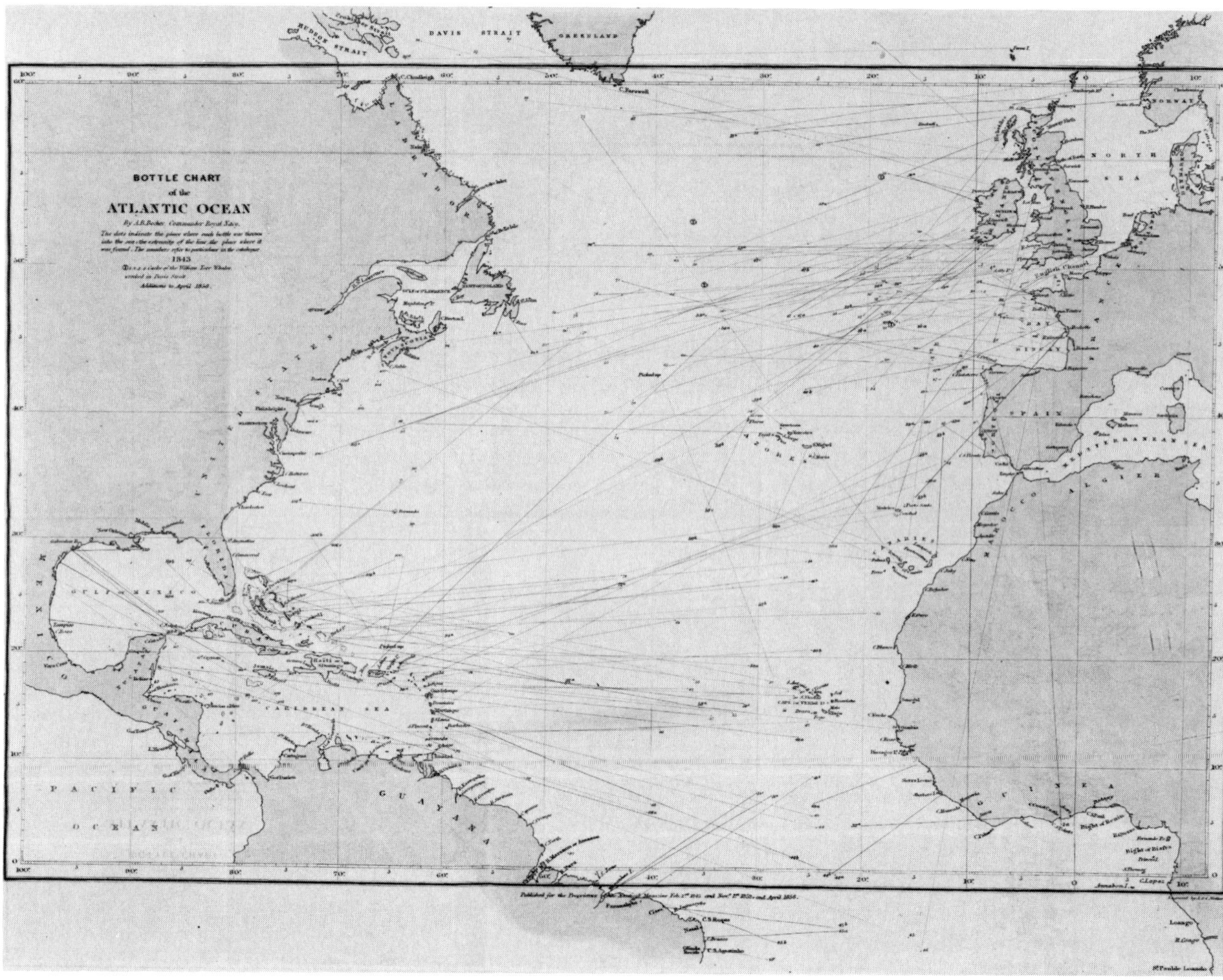

close to a vessel it is possible to arrive at a rough estimate of their velocity and this, after all, is nothing more than the ancient method of timing log-chips past a moving ship in order to gauge her speed. Beyond land, however, the ship should be anchored, otherwise both float and observer will be subject to the same current and their relative velocities will not be apparent.

At first apparatus for these timed experiments was crude. For instance, to test for shallow currents a float was roped to a sinker and the whole was cast overboard. The float was a barrel, buoy or raft loaded to be barely awash and was thus unaffected by wind, while the sinker was a ballasted barrel or a weighted canvas cone or vane. The proportional areas of float and sinker determined whether a surface or subsurface current directed the overall travel of the assembly. The method was described in 1747 by Bouguer[8] and was used with minor variations throughout the eighteenth and nineteenth centuries. A late example of the current drag is shown in figure 145. The depth of measurement was restricted, however; if the rope joining float and sinker is long then its area exceeds that of the bodies. The whole apparatus then drifts at the average speed of the column of water between the surface and the sinker. To get a successful measurement in deep water a totally different principle is needed.

Georges Aimé, whose researches in the Mediterranean included the study of tides and wave motion, went to great lengths to devise current meters for both direction and velocity. At this period members of the French Academy of Science wanted to know if a submarine current flowed through the Straits of Gibraltar and Aimé set out to answer their enquiries.

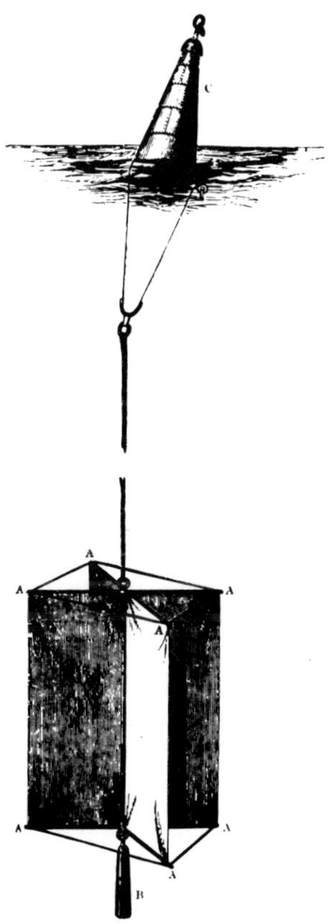

Figure 145. The current drag used on *Challenger*. From *Challenger Report Narrative* 1885 vol. 1, Pt 1, Fig. 25.

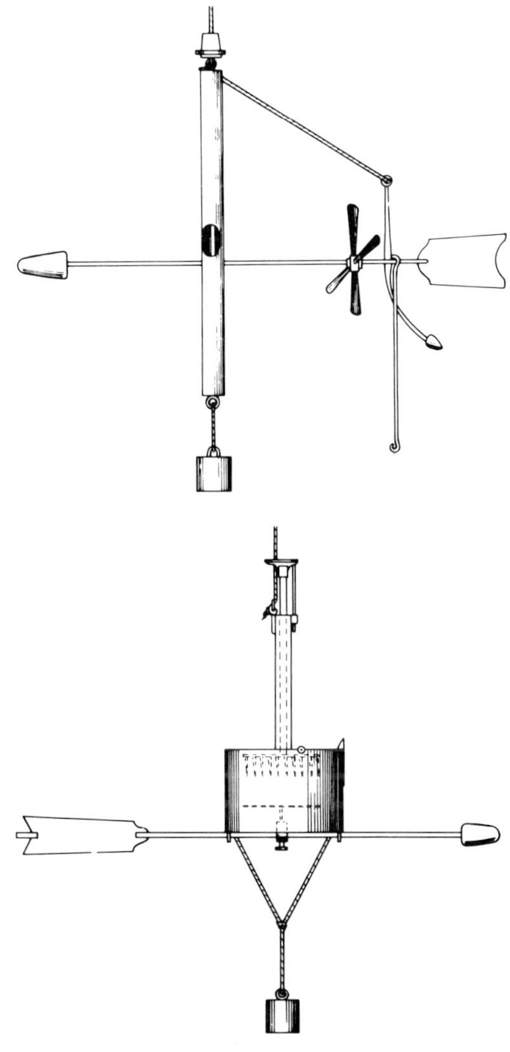

Figure 146. Aimé's designs for current meters for both direction and velocity. Redrawn from Aimé 1843 *Exploration Scientifique de l'Algérie* vol. 1, plate 5.

Previous experiments had been made with a canvas drogue lowered into the sea, measuring the consequent deflection of the towing ship from her course. Aimé designed a true meter but was never able to have it constructed, and so fell back on the 'English method', as he termed it, releasing marked floats from various points. His designs merit description; they were published in his book *Exploration Scientifique de l'Algérie* (1845),[9] and were therefore available to interested readers (figure 146). The direction meter was later constructed and used by Irminger when he investigated the Gulf Stream in the region of Iceland in the 1850s and 1860s.[10]

The direction meter consisted of a toothed wheel sliding over a compass needle, both held in a copper case. Below the case a vane aligned the apparatus in the current. The meter was lowered on a braided silk line since Aimé had found that laid rope tended to unlay (and revolve) in deep water. To register, a messenger was sent down the line to drive the toothed wheel over the compass needle and lock it. When raised, the angle between the needle and vane showed the direction of current flow, relative to magnetic north. Its accuracy was limited to about 6° since the angle subtended between each of the 32 teeth was 11° (a compass needle can be used near a wooden ship, and even below one built of iron or steel, if the compass is lowered sufficiently deep to be unaffected by the ship's magnetic field).

The proposed velocity meter resembled meters used in canals and rivers. A propeller travelling along a threaded rod was held by a restraining linkage as the meter was lowered. A messenger unlocked the propeller, allowing it to rotate and travel along the rod. After a measured time a second messenger locked the propeller and the apparatus was raised. The current velocity was calculated from the distance travelled by the propeller during the elapsed time, with reference to an empirical formula.

11.3. PILLSBURY'S CURRENT METER (1884)

The stimulus given to oceanography by the laying of cables did not extend to current meter work. Cable engineers were interested in the current at the sea bed, but its strength was shown by the nature of bottom sediments. Clay is laid down in quiet water; as the current velocity increases, fine and then coarser material will be winnowed out. The direction and frequency of such currents do not concern the cable engineers who avoid such areas whenever possible. They filed some patents for meters during this period and these often depended on electrical telemetering of the register back to the ship, an unreliable technique, given the problems inherent in handling a composite conducting sounding line.

To obtain accurate results in deep water the first essential was to anchor the ship, as a fixed observing platform. In 1877 a ship was built specially for this purpose by the US Coast Survey; their schooner *Drift* worked between Nantucket and Nova Scotia where she anchored for periods of up to 90 hours in water as deep as 135 fathoms.[11]

The US Coast Survey ran a continuous programme in the Gulf Stream, sounding and taking water samples throughout the water column. In 1883 they decided to test the current at different depths. *Drift* was again used and she anchored in 400 fathoms, using floating cans to measure the upper currents. The following year the Survey Superintendent decided to adapt the *Blake* for deep-water anchoring and to develop a current meter that had been proposed as far back as 1876 by Lt John Pillsbury, at that time executive officer on board the *Blake*.

In order to anchor in very deep water (and *Blake* subsequently achieved this in 2180 fathoms) and to confine the ship to one position, the anchor rope must be kept taut. In this case the ship's motion from waves and weather puts a terrific strain on the rope and its points of attachment. A large accumulator was fitted and special gear installed to take the strain of the rope and transfer it to a downward thrust on the main mast (figure 147).[12]

In his report on the first season's work Pillsbury explained:

> The instrument devised by me to obtain the strength and direction of the currents at any depth consists of a freely-

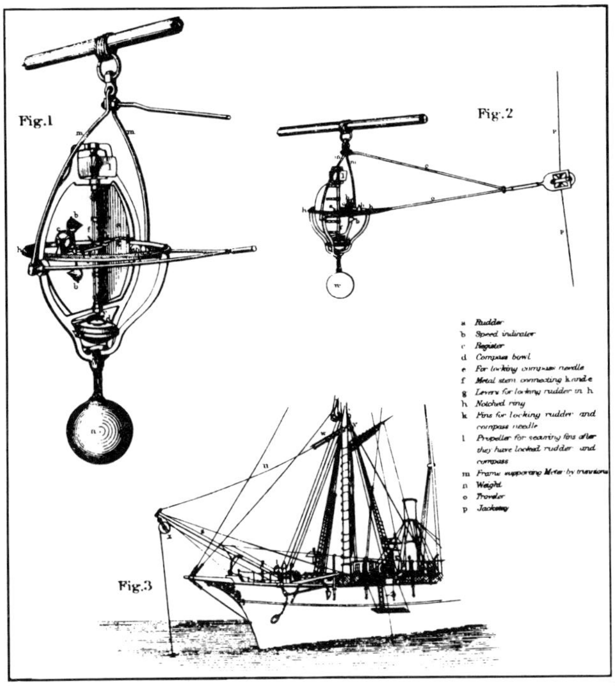

Figure 147. Pillsbury's current meter, as originally made (1876) with the tackle for deep-sea anchoring rigged on USS *Blake*. From Pillsbury 1885 *Rep. USCGS* plate 36.

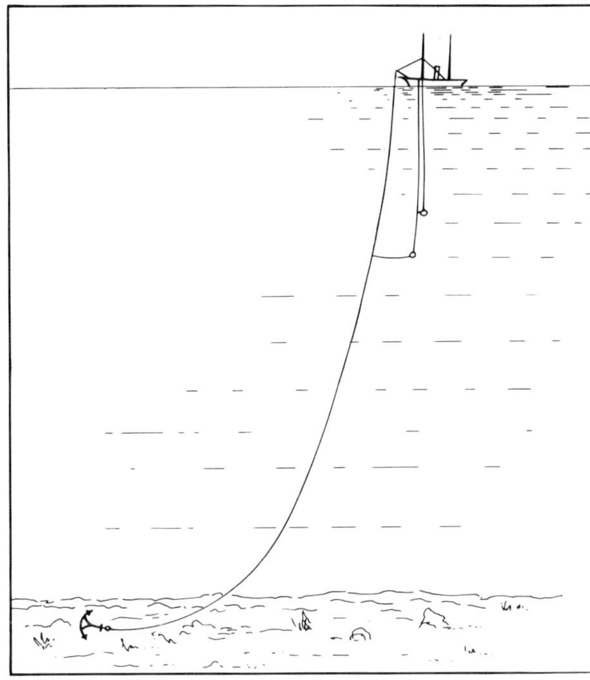

Figure 148. Pillsbury's method of lowering his meter down a jack-stay linked to the anchor cable.

> moving rudder of this metal *a*. On the front of this, and attached to the rudder, is a set of four metal cones *b*, placed at extremities of two arms or spokes which are at right angles to each other . . . At the axis of the system is a small shaft turning in points and by a worm on it turning on it as it revolves differential wheels *c* which register revolutions.[13]

Below the rudder was a compass bowl, with a device to clamp the needle. Once the meter was lowered into the current and aligned by its rudder this needle could be clamped so that, when raised, the meter showed the angle between the current direction and magnetic north. The meter was freed to run in the current for a certain time, then locked and the number of revolutions of the cups (which registered on a counter) used to calculate the velocity of the stream. This locking device proved somewhat unreliable and finally Pillsbury allowed the meter to run continuously while in the water, applying a correction for the time that it was being let down or hauled up. The frame was deliberately made heavy, but in a strong current it could still be deflected out of plumb. In this case its true depth was unknown, and it also failed to register correctly if suspended at an angle. To correct this, Pillsbury rigged a jack-stay, attached to the lower portion of the anchor rope, and sent the meter down this stay by a metal rod-and-roller called a traveller (figure 148). Even so, the two ropes were sometimes swept out in a great bight and on such occasions the depth was found by means of an accompanying Bassnett pressure sounder†.

In 1896–97 the Royal Navy acquired a Pillsbury meter and carried out trials in home and distant waters. Alterations were made, so that the meter could be started and stopped by messenger release (figure 149). The first trials were made in the Straits of Dover, where the meter was run down a jack-stay from HMS *Research*, commanded by Usborne Moore. He found that the messengers appeared to have worked satisfactorily, as the meter came up locked, but having recorded current. However, the results were not those anticipated for in some cases virtually no current had

† Patented as No. 2275 (1879), this was one of a number of navigational sounders available during the late nineteenth and early twentieth centuries, still using the compression-chamber principle described in §6.1.

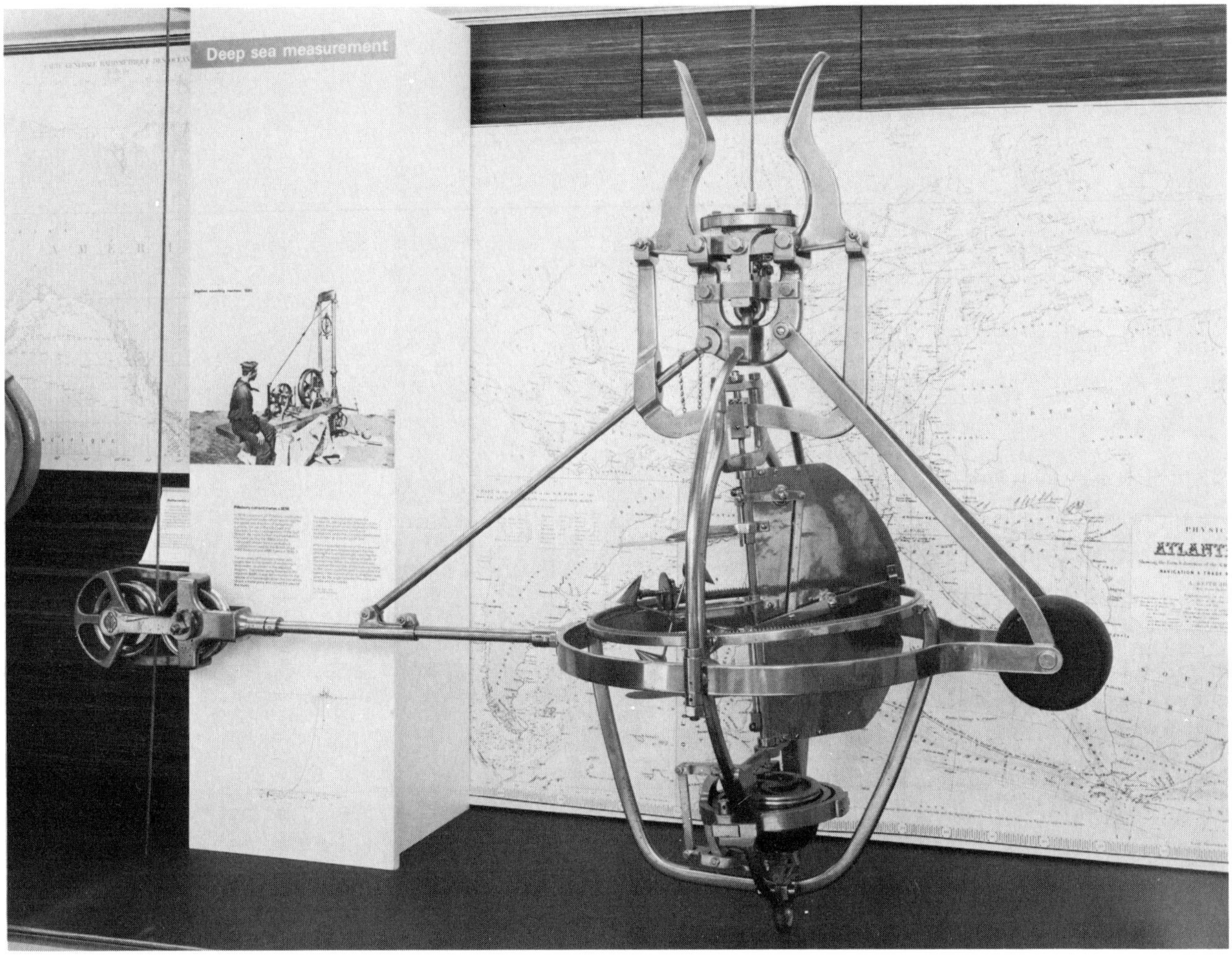

Figure 149. A Pillsbury current meter modified by the Admiralty, 1896–1901.

been recorded in the lower strata when the surface current was flowing strongly, whereas theory demanded that in these tidal waters, surface and bottom currents should run at equal speed. Usborne Moore concluded that the meter had not been hanging plumb in the strong tidal stream (which often reached a speed of 3 knots) and that additional weights were needed. He made the suggestion that:

> All that can be done at present in 20 fathoms where tides are strong, is to make meter observations one hour before and one hour after slack water, and from these to infer what is going on during the strength of the tide.[14]

He had hoped to repeat the observations the following season but the meter was sent instead with HMS *Egeria* to survey the Equatorial Current.

11.4. NANSEN'S PENDULUM CURRENT METER (1900)

Any velocity meter which relies on a windmill arrangement of cups or vanes is ineffective at slow current speeds because it fails to align itself in the direction of flow and water simply moves past the meter without turning the cups preferentially in either direction, if indeed at all. Nansen, trying as always to develop and test instruments to the limit of their capabilities, needed a meter to detect the slow motion of deep water, both the constant flow and tidal oscillations. He wanted a meter which could be suspended from a stable platform, such as an ice-floe, or which could be set in a frame on the sea bed. His solution was to design a sensor which was weightless in sea water and responsive to the slightest water movement and which would register the direction and extent of movement once the meter was in equilibrium with its surroundings. The first experimental pendulum current meter was described by Ekman (figure 150) thus:

> The current is measured by the displacement of a pendulum hanging freely by a thread, the pendulum being arrested *in situ* by a messenger simultaneously with a magnetic needle, which subsequently indicates the direction of the displacement. To the pendulum, which is furnished with six wings acting as surfaces of resistance to the current, is also attached a hollow glass float of sufficient buoyance to keep the pendulum afloat.[15]

The lower end of the pendulum carried a point; below this was a cork-covered plate which, in response to a messenger striking the frame, was levered up to receive the impact of the point which thus registered a certain distance and direction from the centre of the plate. As the compass needle was clamped at the same instant, the magnetic bearing of the current was easily discovered.

The meter was found to be quite effective on its first trials. Nansen considered that it would be improved by dropping the pendulum onto the registering plate, rather than lifting the latter. He also suggested that the pendulum should be made variable in weight, so that in stronger currents it could be loaded to prevent its being carried beyond the edge of the plate. The apparatus had to be raised and reset after each observation but Ekman claimed that with practice one observation could be made per minute in shallow depths, while five minutes sufficed when working at 50–60 m.

Once the basic design was accepted, the ICES Central Laboratory took over the testing and improvements of Nansen's meter. In his first Annual Report Nansen explained that repeated measurements were now possible.[16] The cork disc was covered with a layer of wax which could be imprinted with a series of measurements, each one triggered by the release of a messenger. Then after the impressions were recorded, the wax was smoothed with a warm knife and the meter lowered for the next series. It worked best when hung in a sturdy tripod resting on the sea bed and was

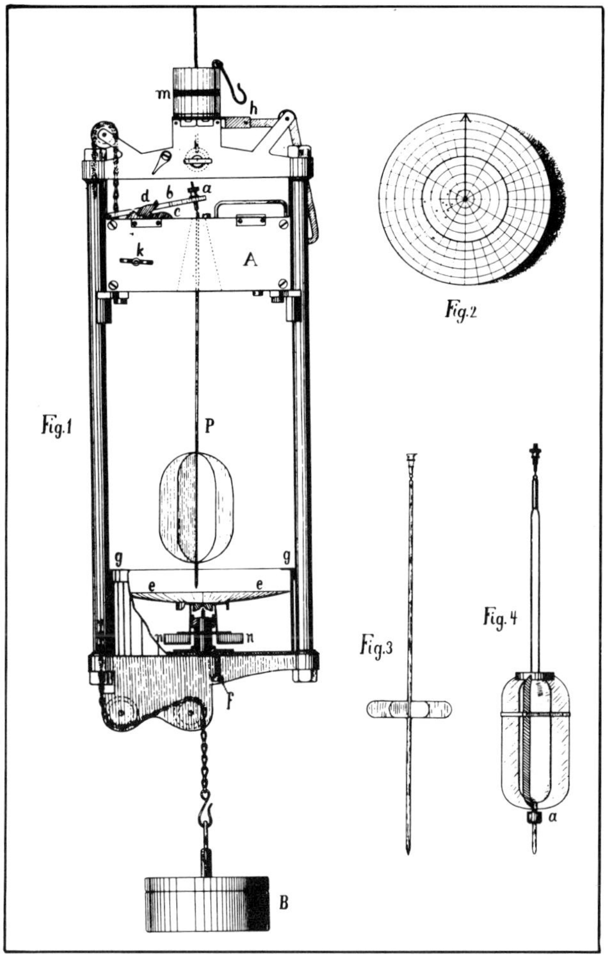

Figure 150. Nansen's pendulum current meter. From Nansen 1906 *Cons. Perm. Int. Expl. Mer Publ. Circ. No.* 34, plate 1.

sensitive to speeds as low as 1 cm/second. The following year further refinements were made, the design was regarded as proven, and the meter and tripod were offered for sale. Mid-water observations, however, were regarded as less satisfactory; when hung from a vessel, even with two anchors out, registration was unreliable if the velocity was less than 10 cm/second.[17]

After the working year 1904–5 Nansen was able to recommend a method of work using a buoy with a clockwork mechanism attached, whereby the meter could be freed, left to operate and then locked, by remote control (figure 151).[18]

Despite the general satisfactory performance of

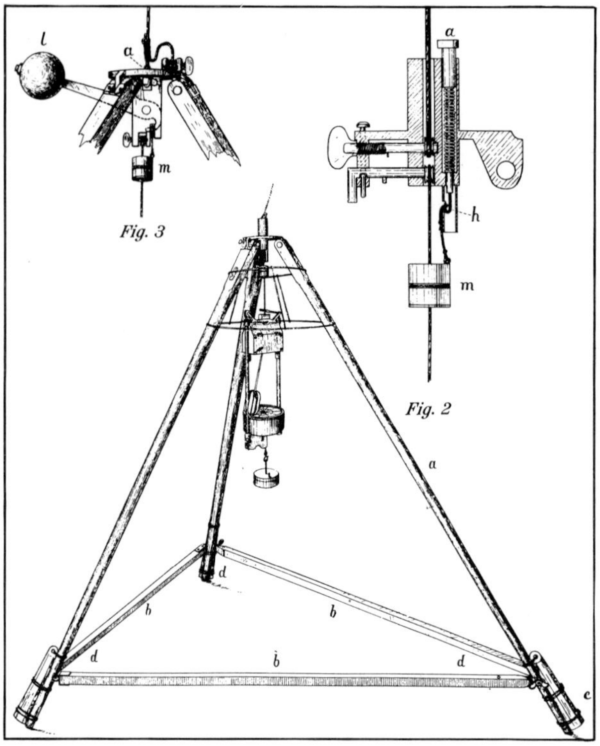

Figure 151. Nansen's pendulum current meter, with clockwork mechanism for repeated registration, 1905. From Nansen 1906 *Cons. Perm. Int. Expl. Mer Publ. Circ. No.* 34, plate 2.

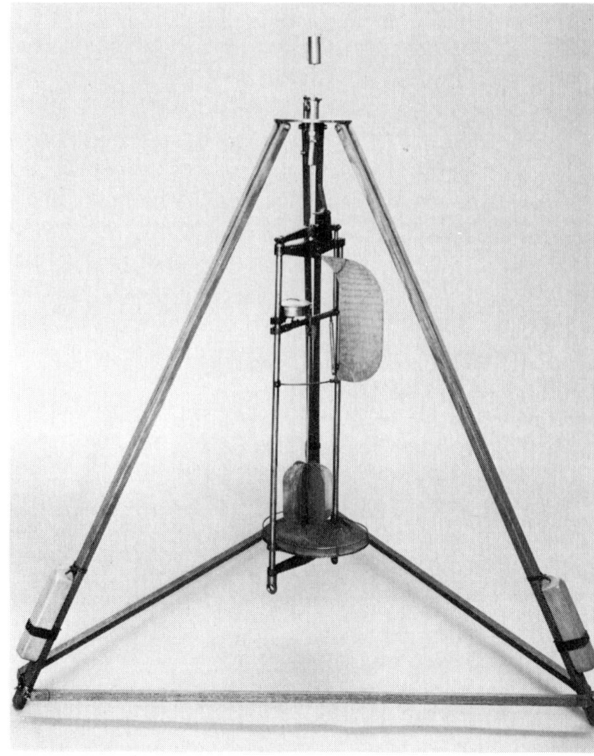

Figure 152. Nansen's pendulum current meter, single-operation type, *c*1905 or earlier.

Nansen's current meter, it was possible for errors to be introduced under certain conditions. The bilateral symmetry of the frame was seen to register a lower force when current flowed parallel with the frame axis than when it flowed at right angles to it. When the apparatus was lowered, there was a risk that it might swing on the end of its line, and also that such impulses might be set up as the messenger was released. In mid-water there was no way of checking these movements; as a result, the meter worked best at the sea bed. Various three- and four-limbed frames were devised, to take account of the depth and nature of the bottom, and the anticipated current strength. The clockwork mechanism was triggered either by messenger from the ship or by the arrival of the frame at the bottom. Figure 152 shows a single-operation meter, with a modified pendulum, operated by messenger release. This design would appear to predate the 1905 clockwork-driven meters.

11.5. EKMAN'S CURRENT METER (1902–3)

The Pillsbury meter was deliberately made heavy so that the instrument itself would not be deflected by the current that it was to measure. An instrument was required to fill the gap between the Pillsbury model and the lightweight velocity meters such as those of Makaroff and Arwidsson (figures 153 and 154) which were to be hung a few fathoms below their controlling ships.

ICES Laboratory staff concluded that for regular use at sea, both propeller and pendulum meters were essential if the entire range of velocities was to be measured. Nansen's meter was conceived before the Laboratory came into existence, but Ekman's propeller meter had been invented and tested out on the *Michael Sars* during the first year's work,[19] together with a means of anchoring it at the bottom so that it was unaffected by waves and strong surface currents. The following year (1904) it was available for purchase, its design having been finalised, although experiments continued in the hope of finding a way of operating it independently of the parent ship. In 1905 it was the subject of a special publication by Ekman who described its construction (figure 155).[20]

The meter hung in its frame in ball bearings and was turned into the current by a light tail fin. Once it was set running by messenger release the direction which it had taken up was registered in a most ingenious manner. A compass needle with a grooved upper surface was mounted below the gear train and a hopper, regulated by the propeller, dropped lead shot at each 50 revolutions. The shot ran along the compass needle and fell into one of 36 compartments forming the compass

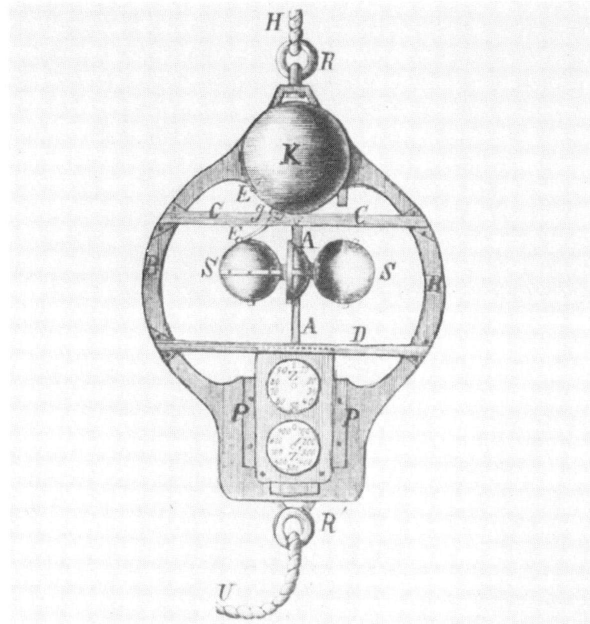

Figure 154. Arwidsson's current meter. From *Berlin Fischerei-Ausstellung* 1880, Fig. 24.

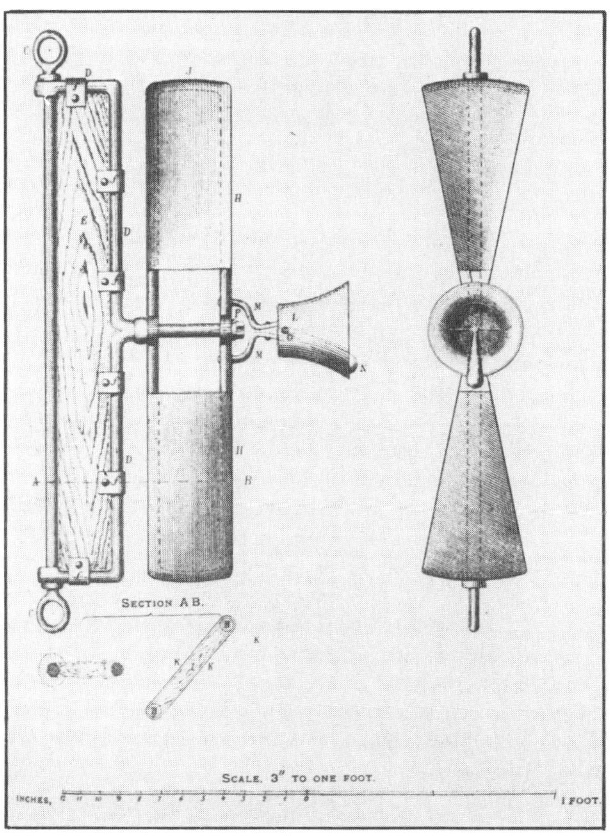

Figure 153. Makaroff's current meter. From Makaroff 1899 *Proc. R. Soc. Edinburgh* **22** plate 2.

bowl. The meter thus not only indicated the direction that it faced but also the extent of any deviations; that is to say, it showed whether the current was constant or if it veered about a mean. After a measured time interval a second messenger was sent down to lock the mechanism.

In 1906 Ekman added a refinement by replacing the lead shot by numbered bronze balls, and by redesigning the hopper and feed mechanism he arranged for three balls to fall into the compass box for each 100 revolutions of the propeller. In this way he could discover the characteristic movements of a body of water at each state of the tide or at different seasons. This meter was regularly used on the *Michael Sars* during subsequent years and found favour on many research ships as the first reliable means of measuring currents. The apparatus underwent repeated modifications to suit it to various conditions of use, and it continues to give service at the present time.

From the ICES Laboratory came notice of other types of meter, for example, those of Witting, Jacobsen and Pettersson were described in the Commission's *Occasional Papers* series. However, these instruments were intended for near-surface work and involved the use of electrical telemetering (those of Witting and Pettersson) or direct observation of the angle of line, in

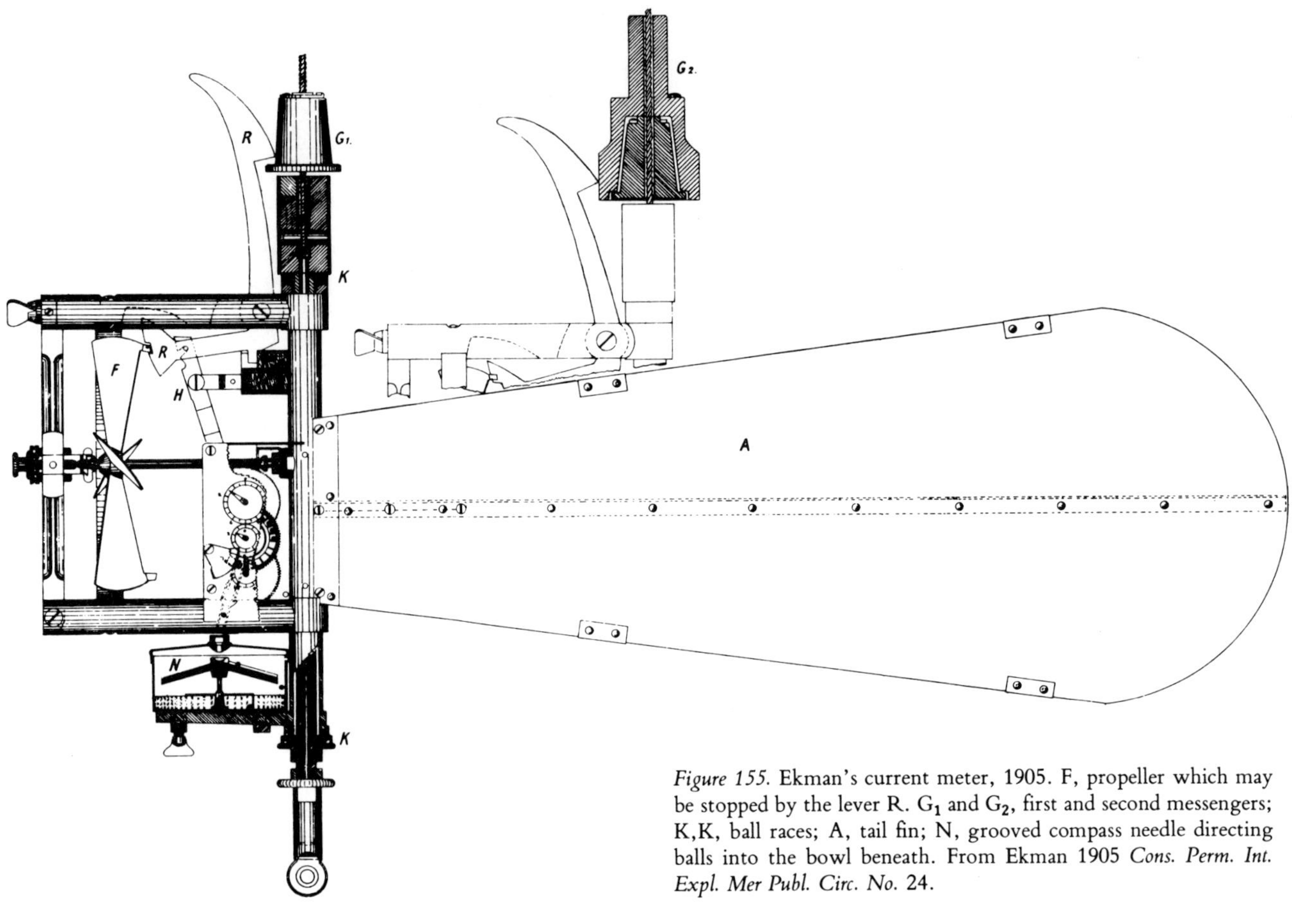

Figure 155. Ekman's current meter, 1905. F, propeller which may be stopped by the lever R. G_1 and G_2, first and second messengers; K,K, ball races; A, tail fin; N, grooved compass needle directing balls into the bowl beneath. From Ekman 1905 *Cons. Perm. Int. Expl. Mer Publ. Circ. No.* 24.

effect a pendulum meter suspended from the ship (that of Jacobsen). This latter was intended to show the direction of a current so close to the hull that a magnetic compass would not be appropriate. These meters were often used from stationary vessels such as lightships, from which continuous records of current movement throughout the tidal cycle could be obtained.

REFERENCES AND NOTES

1. Bohnecke G 1954 'The principles of measuring currents' *Ass. Ocean. Phys. Publ. Sci. No.* 14
2. Becher A B 1857 *Naut. Mag.* **27** (unnumbered folded chart)
3. Redfield W C 1845 'On the drift ice and currents of the North Atlantic' *Silliman's Journal* **48** pp373–88
4. Aimé G 1845 *Exploration Scientifique de l'Algérie 1840–42* (Paris) vol. 1, pp186–91
5. Mitchell H 1859 'Report of Assistant Henry Mitchell on the physical surveys of New York Harbor and the Long Island with descriptions of apparatus for observing currents etc' *Rep. US Coast Survey* Appendix 26, pp310–17
6. Richard J 1900 *Les Compagnes Scientifiques de SAS le Prince Albert 1er de Monaco* (Monaco) pp51–6
7. Mitchell H 1859 *Op. cit.* pp314–5
8. Bouguer P 1747 'Sur une nouvelle construction de loch' *Hist. Acad. R. Sci. Paris* **60** pp96–102; and *Mém. Acad. R. Sci. Paris* **60** pp646–64 (see also §4.2)
9. Aimé G 1845 *Op. cit.* pp181–5
10. Berlin: Fischerei-Ausstellung Wissenschaftliche Abtheilung 1880: *Instrumente zur Untersuchung des Wassers* (Berlin) p19
11. Pillsbury J 1889 'The Gulf Stream—a description of the methods employed in the investigation, and the results of the research' *Rep. USCGS* Appendix 10, p512

12. Pillsbury J 1885 'Report on deep-sea current work in the Gulf Stream' *Rep. USCGS* Appendix 14, p496
13. Pillsbury J 1885 *Op. cit.* p497
14. Usborne Moore W 1899 *Report on Observations of the Tidal Currents and Undercurrents in the Strait of Dover made with a Deep-Sea Current Meter . . .* (London) p1
15. Ekman V W 1901 'On a new current meter invented by Prof. Fridtjof Nansen' *Nyt. Mag. Naturvidernskaberne* **39** pp164–5
16. Nansen F 1903 'Report of the Central Laboratory . . . 1902–3' *Cons. Perm. Int. Exp. Mer Rapports et Procès-verbaux* pXLIII
17. Nansen F 1904 'Report of the Central Laboratory . . . 1903–4' *Cons. Perm. Int. Exp. Mer Rapports et Procès-verbaux* pXXI
18. The report explained how this was achieved (see the series referred to in refs (16) and (17) ppXXIII–XXIV) and instructions were repeated in the *ICES Publ. Circ. No.* 34 (1906) Appendix by Ekman pp33–4
19. Nansen F 1903 *Op. cit.* pXLIII
20. Ekman V W 1905 'Kurze Beschreibung eines Propell-strommessers' *Cons. Prem. Int. Exp. Mer Publ. Circ. No.* 24

Chapter 12

Factors Controlling Technological Progress in Instrumentation

12.1. POLITICAL AND COMMERCIAL STIMULI

Throughout the time spanned by this account, specific instruments were invented or markedly improved by individuals. But oceanography is not advanced by single waterbottles, sounders or thermometers, however efficient they may be. The mode of use is equally important: the gear by which they are lowered into the invisible depths, and the ability of the ship to pinpoint its location upon a chart. Whereas earlier instruments were designed by scientists or inventors who were not naval men, their deployment was in the hands of private captains or the navies of one nation or another. Only at the end of the nineteenth century do we see the rise of an organisation whose team of scientists design, test and use an instrument and themselves interpret its results. ICES is the prime example of such teamwork. Earlier successes were often due to the same mode of operation, albeit on a limited, because private, scale. The work of Marsigli, Scoresby, Aimé and perhaps Prince Albert of Monaco fall into this class, though the latter occupies a position approaching that of an entire navy.

It is contended that *significant* improvements in underwater apparatus came at those times when: (*a*) a navy considered that its surface operations would be thereby advanced; or (*b*) commercial interests demanded it. These improvements depended for their execution on: (*c*) the availability of suitable materials, including ships and technological processes, and (*d*) adequate manpower to work intensively in the chosen region. Whereas (*a*) and (*b*) are alternatives, (*c*) and (*d*) are essential factors.

In the early nineteenth century the British navy tried to summon oceanography to its aid when instructing its captains bound for the Arctic regions. As stated explicitly, the Admiralty hoped that deep-water investigations would reveal a route through ice- and fog-bound polar seas where visual sightings and common navigational aids were of little use. The necessary backing of ships and crews was provided because there was a strong incentive, in terms of geographical discovery. The failure to advance marine science was, in the author's opinion, due to the lack of suitable materials for deep-sea instruments.

British and American navies were requested to support cable-laying operations because both governments saw considerable advantage in drawing tighter the lines of communication with Empire or trading partners. Strong central government control and faster reaction times were essential for Britain to hold together her far-flung realm. However, naval vessels were unlikely to profit by sounding expeditions except in those areas soon to become theatres of war, such as the Mediterranean and Black Seas. Interest in the eastern Mediterranean was heightened by the impending disintegration of the Ottoman Empire and the decision to build the Suez Canal.

The zoological research cruises, undertaken over the same period as the cable surveys, appear to have been no more to the British Navy's liking; HMS *Lightning* was provided for one of these because, as the oldest steamship in the fleet, she was useless for any other purpose. In Britain these cruises, and the cable-laying exploits, captured the attention of a public avid for new discoveries, and they were written up in many popular magazines. In North America the annual cruises of *Blake* were continued by vessels of the US Fish Commission and there was a greater willingness to experiment with new apparatus, encouraged perhaps by the proximity of the area studied to the ships' home base with its engineering facilities. American scientists suffered a setback during the divisive Civil War in 1862–65, without which they might well have gained and held a technical lead over Britain.

Commercial interest in oceanography, with its ability to raise capital, was a product of the Industrial Revolution. As the British government nationalised the inland cable network, releasing a large quantity of money for future investment, the cable companies took over the business of making preliminary surveys and from the 1870s it seems probable that they surveyed annually more route-miles than did the Navy. The difference in competence between naval and commercial vessels was explained by Rear-Admiral Wharton, then Hydrographer, in 1896 at the Colonial Office inquiry into the matter of laying and operating, on behalf of the government, a transpacific cable from Vancouver to Australia. It was the first cable that would have to pass near, and perhaps cross, the newly discovered zone of deep ocean trenches. Questions arose as to the length of time required to survey the route and whether this should be done by the cable company selected or by the British Navy. Wharton gave the following evidence:

> *Wharton*: At 3000 fathoms soundings take about 2½ hours in our small surveying vessels . . . the deep soundings . . . can be done quicker and more readily by the cable ships.
> *Chairman*: In fact the navy are not provided with such perfect appliances, or ships so well furnished for this particular purpose, as those whose business it is to do it and nothing else?
> *Wharton*: Yes, that is so, and besides the reason that they are wanted for other purposes. That is the thing.[1]

The effectiveness of commercial surveys was summarised by Peake in 1901 with reference to two transatlantic surveys that his company had recently carried out. Soundings in 3000 fathoms took only 1½ hours. Closely spaced soundings along a zig-zag route gave the contours of the area, as opposed to a profile. Many new topographical features were recognised for the first time—after 40 years of surveys across the Atlantic—volcanic mounts were found to be numerous but frequently so rocky that no material was collected from the bottom, evidence being in the form of a battered sampler tube. On one of these expeditions the Telcon ship *Britannia* left England on 4 May 1899 and returned on 3 August, during which time she steamed 6678 miles, taking 477 soundings, 150 temperature measurements and some current meter readings between the Azores and North America.[2]

In the last period covered, that which saw the establishment of cooperative fisheries research, there was considerable advancement in the design of apparatus and here there was a commercial interest, one represented by governments of several nations acting in concert. It would therefore appear that conditions (*a*) and (*b*) were favourable to the improvement of apparatus: (1) in the early part of the nineteenth century; (2) in the last quarter of the nineteenth century, for sounding; and (3) during the period 1890–1914. Such potential will only be realised if conditions (*c*) and (*d*) are fulfilled.

12.2. TECHNICAL COMPETENCE

Technical competence contributes to improved design or construction of instruments as for example by fitting better valves or more precise gear mechanisms. New materials, or those not previously chosen, may be introduced. The substitution of brass and glass for the wood and organic materials of the early sounders and waterbottles was as important as the eventual acceptance of ebonite and celluloid. The apparatus may be more effective if it is lowered from a steamship able to keep on station during the operation, by new methods of handling gear and by the foresight of the expedition commander who has proposed a suitable programme of investigation. With these points in mind, the various classes of instrument will now be inspected more closely.

Water samplers were markedly improved by metal construction (Scoresby 1817, Davy 1818, Marcet 1819) and by the adoption of positive closing mechanisms.

The messenger-operated valved bottle recommended by Marcet is known to have worked to some effect, whereas the earliest propeller-closed devices (Smithson Tennant 1811, Ross 1818) were unsatisfactory and did not survive.

The continuing use of free-moving valves (Admiralty standard types pre-1872); bottles with a single opening which perforce went down full of air (Ross 1818, Stevenson 1830), plunger forms (Biot 1838, Rung 1886), gave way to two new designs: the stopcock bottle (Buchanan 1872) and the slipping-cylinder bottle (Ekman 1869, Meyer 1872).

Apart from changes of form, new materials were incorporated: glass (Biot 1838, Nansen 1900), gutta-percha and rubber (Ekman 1869, *Travailleur* 1883), ebonite and celluloid (Pettersson 1890, Automatic 1903). Flap-valves eventually returned, but with spring closure. These valves always had the advantage of allowing water to flow freely through the bottle on its way down since they could be made as large as the circumference of the bottle if both opened outwards, which was possible only with spring closing.

Although favoured by Cavendish, Aimé and Walferdin, the overflow thermometer never gained the popularity of the Six instrument which served on ships of several nations for a century or more. During the course of its life the thermometer was given hair-spring indexes, a sealed tube, a creosote–alcohol fluid (all these improvements predating 1850), a pressure shield round the bulb, reduced size, and a change to two-limbed form. A thermometer with a floating index is necessarily of wide bore and for a given change in temperature the mercury will not move far up or down the tube unless the bulb is very capacious; hence it was never possible to read these instruments with accuracy. On *Porcupine* scientists recorded to 0.01 °C; on *Challenger* to 0.01 °F, but the readings often seem to have been made only to ¼ °F and indeed smaller fractions were subsequently shown to be invalid. To the end of its life the Six thermometer was beset by the problems of shifting indexes, the need for careful handling and transport in a vertical position, and the tube's vulnerability to compression. Its greatest drawback was its inability to locate the extremes of temperature registered.

The reversing thermometer (Negretti and Zambra 1874) did accomplish this, and was wholly protected inside its glass sheath. The tube was drawn so fine that fractions of a degree could be engraved on the stem and the accuracy of manufacture ensured that the mercury thread broke off only at the constriction. Capsizing-thermometer frames of various patterns operated the thermometer reliably, at the intended depth. Thermometers differed from other physical oceanographic instruments in that they could not be made or modified on board ship, but they were always made and calibrated by specialist instrument-makers.

Lineless sounders were doomed to failure and never replaced the deep-sea lead and line. The tendency toward lighter gear (Scoresby 1820) rather than the great overall weight favoured by the Arctic explorers (Ross 1818, Parry 1819–23) led eventually to the use of silk, twine (Maury 1850, Spratt 1855) and finally wire line (Thomson 1872), with detaching-weight sounders (Brooke 1852).

Waywiser sounders benefited from the development of ships' logs during the late eighteenth and early nineteenth centuries (Foxon 1772, Gould 1799, Massey 1802), while Massey's air tube made it possible to obtain accurate measurements irrespective of the sounder's rate of descent. By the early nineteenth century also, tools for accurate gear-cutting and machining were to hand in the instrument-maker's workshop.

Glass was an ideal material for pressure-sounders for it allowed an indicating fluid to be read directly (Hales 1728, Ericsson 1835), and this effect was enhanced by chemical registration (Thomson 1880). Further improvement was limited but these sounders conferred the ability to sound with the ship under way and survived well into the twentieth century as navigational aids.

The sounding machine, from its tentative origins in the simple raft or buoy and nipper (Massey 1802, Burt 1818), the spring balance employed by Spratt (1854) and the early abortive attempts to sound with copper or iron wire, had to await the production of spliced lengths of high-tensile wire (Thomson 1872), steam power, and strong accumulators (Sigsbee 1874) before the various components were brought together as a self-contained machine.

The steam engine was introduced gradually into seagoing vessels from the middle of the nineteenth century. It served to propel the ship and to power auxiliary machinery. As the ship's range was limited by the tonnage of coal that had to be carried and the location of shore coaling stations, steam was not universally welcomed. Well into the century ships continued to be built, as *Challenger* was, to move under sail or steam, and this greatly restricted the space available both above and below decks. The purpose-built cable ships, which did not as a rule leave commercial routes, were able to rely on steam power alone and could remain on station and work in most weathers.

Although current meters based on waywiser, paddle-wheel, Venturi tube and pressure plate principles were constructed for river and canal gauging, deep-sea currents were not measured until a means of anchoring in deep water was perfected (*Drift* 1883). Then, revolving-cup meters (Pillsbury 1884) and pendulum meters, the equivalent of the pressure plate (Nansen 1900), were successfully used in fast and slow waters respectively. Non-magnetic and non-corrosive metals were essential for bearings and accurate machined parts.

The provision of adequate manpower to work intensively in the chosen area was contributed by various navies in times of peace when, following the recent cessation of a war, fleets were up to strength and trained crews were available. In Britain the number of ships decreased when no danger threatened, the government being reluctant to allocate money for shipbuilding at such times. In the early nineteenth century, after the Napoleonic Wars, the population of Britain increased rapidly despite emigration, ships and men were available, and from then on commercial enterprises had little difficulty in recruiting skilled labour.

Given three periods when improvements might be expected since the conditions (*a*) and (*b*) above were fulfilled (see §12.1), it seems that oceanography did not progress far during the early nineteenth century for want of adequate technology. The new instruments were simply not competent for their tasks, they were hampered by the methods of work and their results were not easily interpreted since cruise schedules did not allow return for checking anomalies; indeed, the practice of making circumnavigations militated against a successful oceanography programme. Further, oceanography, though acknowledged to be important, still took second place to the business of exploration, rather than leading it onward.

As for the scientific work accomplished in Britain during the latter years of the nineteenth century, scientists gained their reputations mainly in the field of zoology, the result of their dredging programmes and accomplished despite, rather than as a result of, the physical oceanographic apparatus with which they were provided. There was no equivalent to the International Laboratory in which scientists designed good apparatus and then accompanied it to sea for trials and eventual use. Reluctantly, it can be concluded that during the nineteenth century, in Britain at least, the aims of science were not assisted to the extent that the aims of commerce were. This was made clear when the British Association for the Advancement of Science came into being and, despite their efforts, and those of the Royal Society, it has continued to be the case. It was perhaps in keeping with the strong commercial drive and the philosophical spirit of the age, but it is regrettable that the maritime kingdom of Britain, though it had a King, George III, who was interested in astronomy, never had one such as Prince Albert of Monaco who stimulated every aspect of oceanographic research in the Mediterranean and adjacent seas.

REFERENCES AND NOTES

1. GB Colonial Office 1899 *Report of the Pacific Cable Committee* (London) pp76–80.
2. Peake R E 1901 *On the Results of a Deep-sea Sounding Expedition in the North Atlantic During the Summer of 1899* (London)

Bibliographic Study

The sources of information concerning instruments and apparatus prove to be very varied. For the earliest period—up to 1800—there are a few books and encyclopedia articles, but most information has been drawn from journals of the academic and philosophical bodies then active in London and Paris. Despite these publications it seems that little effort was made to perfect the apparatus for use. It should be remembered that during the eighteenth century ships were still small; even the East India Company's vessels, considered to be among the finest afloat, were no more than 450 tons and these carried crews of only 90 men, divided into watches. To haul up a deep-sea line required a capstan manned by 20 or more sailors, a number not easily spared from their normal duties. Royal Navy vessels always carried a larger crew, commonly twice as many men, these being required to man the ship's guns. During time of war large numbers of soldiers or marines were on board and might be available to supply extra manpower. The Royal Navy also had the logistic ability to make, store and issue apparatus and to require its use and a report on the results. Since the Royal Society enjoyed Royal patronage it was possible for the members to take advantage of that fact to secure assistance from naval ships in carrying out their experiments. Recall, however, that the Hydrographic Department of the Navy was not established until 1795, so that no clearly defined programme of exploration existed. Nor were commercial vessels in a position to stand-to and undertake experiments to gratify the curiosity of the captain, though this might be done if the ship were becalmed, or if specific requests had been made by the owner.

The extent to which encyclopedia editors had to turn back to much earlier work in order to describe the sea is made clear in their articles. *The Complete Dictionary of Arts and Sciences,* published in 1773, carried under the heading 'Sea' an article which drew heavily on Marsigli's book, to which it gave due reference and credit. A second article on 'Current' described ways of measuring surface and submarine currents with reference to the ship's position by lowering a heavy weight or sea anchor beneath the surface.

Rees's Cyclopaedia of 1819 also carried a long article about the sea, drawn partly from the older work of Robert Boyle. It did however refer to the Hales 'bucket' which it categorised as similar to that invented by Hooke, and to the Hales 'Sea-Gage'. It is the use of an engraving of Hooke's bucket to illustrate this article that leads the author to believe that no engraving was ever made of the Hales bucket.

As late as 1823, when ships of the Royal Navy and from other nations had begun to explore certain regions of the oceans, the *Encyclopaedia Britannica* of 1823, under the entry 'Sea', had to report:

> Through want of instruments the sea beyond a certain depth has been found unfathomable.

The article went on to discuss the depths that probably existed. Assuming that the sea mirrored the land in elevation, it would not exceed five or six miles. However, the 1842 edition of the *Britannica* suggested that the depth might be proportional to the extent and in this case might be up to 15 miles deep.

Information and discussion became more readily available to the curious reader as the nineteenth century dawned. Those works accessible to the general public, as opposed to naval logbooks and records, may be divided into three classes: books, periodical literature and patents. The books, as shown by many of the references used in this study, were mostly in the form of narratives of voyages. Those written by expedition commanders were based on the journals which they were required to hand over to the Admiralty on their return. Published with commendable swiftness, they often appeared in the year following the voyage, and had a considerable scientific content, frequently including *Instructions, List of Apparatus* issued, *Report* on that apparatus, *Tables of Results* and various appendixes on aspects of a particular science. The methods used were often described within the narrative and were occasionally illustrated, as in the case of James Clark Ross's Antarctic deep sounding (figure 42) and Scoresby's waterbottle (figure 35).

Works of a more general nature, such as those written by junior members of expeditions for popular consumption, are on the whole irrelevant here as they give conclusions without explaining the method of research and also, perhaps with an eye to future sales, they devote more space to the exotic plants and animals and to the curious behaviour of the native peoples encountered, than to the repetitive tedium of deep-sea sounding operations. However, good scientific writing is by no means restricted to naval officers for the two books by Scoresby, summarising his Arctic researches as a whole, are of great interest to the student of natural history, meteorology and oceanography.

The earliest periodicals from which information has been drawn for this study are the *Philosophical Transactions*, published privately to begin with, by Henry Oldenburg, Secretary to the Royal Society, and the *Journal des Scavans*, both of which began publication in 1665. Production of the French journal became somewhat erratic around the turn of the century and it was always less concerned with science than its London counterpart, though both published abridged translations from each other's papers. The *Philosophical Transactions* eventually became the official journal of the Royal Society and continues so to the present day. The *Gentleman's Magazine*, founded in 1731, published articles and letters from Hales and from Six but during the nineteenth century did not concern itself with marine science.

The improvement to roads and postal services during the late-eighteenth century allowed the ready circulation of a growing number of periodicals which began to appear in many European countries around the beginning of the nineteenth century. There was a dramatic increase in the number of societies in existence and details of apparatus appeared in journals devoted to chemistry, physics, natural history, mechanical arts, naval science and science in general, as well as continuing to figure in encyclopedias. The periodical literature may be divided into four classes: *journals; transactions* and *proceedings* of learned societies; *abstract* and *review journals*, and *encyclopedias.*

Abstracts and reviews did appear in all the first three types of periodical literature mentioned above, often without naming the original sources. With the rising population of literate mechanics and craftsmen there appeared technical journals which carried some original articles, but obtained much of their source material by reviewing other journals and abstracting patents and notices of inventions. The dates when some of the relevant periodicals were founded is of interest for the light shed on the increased transmission of knowledge through workers in trade and industry, as opposed to the academic scientist.

1794 *Repertory of Arts and Manufactures* (became the *Repertory of Patent Inventions* in 1825)
1797 *Journal of Natural Philosophy (Nicholson's)*
1797 *Philosophical Magazine*
1813 *Annals of Philosophy*
1818 *Edinburgh Philosophical Journal*
1819 *Journal of Science and Arts* (*Silliman's*, published in America)
1820 *Journal of Arts and Sciences*
1823 *Edinburgh Journal of Philosophy*
1823 *Mechanic's Magazine*
1832 *Nautical Magazine*

The encyclopedias have been grouped with the periodical literature because some of them went through several editions, and efforts were made to update their contents. Outpaced by the growing number of magazines which appeared monthly, or even weekly, the information presented by encyclopedias became very out-of-date. They also tended to use old blocks for printing the illustrations and are therefore unreliable as source material.

Patents are of great interest as they reveal ways in which various people, not simply instrument-makers, were trying to solve the problem of gauging water flow. The sounder is the only type of oceanographic instrument to figure in patents up to the middle of the nineteenth century. Neither waterbottles nor marine thermometers are mentioned and current meters are referred to only in so far as the same mechanism might serve to gauge the relative speeds of ship and surrounding water. It is possible to trace the evolution of the waywiser sounder and its associated mechanism by looking, as we have seen, at the patents for logs that were granted during the late eighteenth and early nineteenth centuries.

There are two possible reasons for the increased number of sounder patents taken out during the turn of the nineteenth century—the first is that some of the instrument-makers concerned, Massey and his family among them, regularly patented all their inventions, and the second reason may lie in the similarity between logs and sounders. Ship's logs were required in greater numbers than sounders and any novel invention commanding a good market might be pirated by another manufacturer and the sales of the original inventor thereby reduced. In contrast, few persons wished to make deep-sea soundings other than from vessels equipped by HM Dockyards. It seems not to have been the practice for Naval officers to patent their inventions when these were destined for use aboard Royal Navy ships. Increasing use of shallow-water (or navigational) waywiser sounders through the early years of the nineteenth century is attested by the space which journals gave to publicising such mechanical inventions and by the bitter feuds which protagonists fought over rights of invention when future sales were at stake.

The earliest patent which refers to sounding is one granted to Sir John Christopher van Berg in 1636 (*GB Patent Specification* No. 92a). However, an inspection of the full text reveals that it is a general licence for

> . . . The sole makings and vse of the said Invencõns, engynes and instruments . . .

The list contains a whole variety of mechanical devices said to have been observed by van Berg in 'other countries', among them

> . . . instruments of soundings and fathomings of any depths, whether of waters or mynes.

A similar patent is that granted to George Reynoldson in 1731 (*GB Patent Specification* No. 534), relating to

> . . . a machine that gives exactly the way which a ship makes, and counts her leeway, that tells what speeds the winds blow, and proves the different force of current in any depth . . .

Both these early documents are of the legal type in which Letters Patent confer a monopoly to manufacture. No specification is required or given. Such monopoly of manufacture was intended to encourage the import into Britain of useful inventions from abroad, ensuring some return on the investment made in setting up workshops or manufactories and employing craftsmen in this country. The reason that van Berg's patent is often quoted as a 'first' in the field is because Bennet Woodcroft included it in his *Abridgements of Patents* (1636–1866) under class 76, 'Philosophical Instruments'. An inspection of the entire document, rather than the abridgement, however, reveals its true nature.

The now-familiar form of patent, incorporating a specification, appeared late in the eighteenth century and the early examples which dealt with waywiser logs have been discussed in §3.3. Several instruments were patented in Britain and America simultaneously, others on one side or other of the Atlantic alone. The status given to any article by the granting of a patent is made clear by the frequency with which the word 'Patented' is engraved on such articles, often without benefit of patent number (sometimes, probably, without benefit of patent granted in either country). In the second half of the nineteenth century about 100 patents were granted for sounders intended for medium- or deep-water use. Additionally, patents can be found covering marine thermometers (notably the reversing thermometer patented by Negretti and Zambra in 1874 (*GB Patent Specification* No. 109). Other instruments were patented either in their entirety or in some part which might be used also in thermometers destined for other specialist uses. As with the sounders, the granting of a patent was no guarantee that an instrument could be made to work successfully, and many instruments figured in the index of patents are never heard of again.

After the middle of the nineteenth century another three new classes of information appear: the illustrated instrument-makers' catalogues; the often-illustrated catalogues of the great international exhibitions so popular during the second half of the century; and the published reports of Parliamentary commissions set up to enquire into the state of the nation's fisheries and into aspects of submarine cable-laying.

The instrument-makers' catalogues ranged between the books of several hundred pages put out by Negretti and Zambra (1859, 1864, 1880, 1885), through the more modest catalogues of Casella, Elliott, and Pastorelli and Rapkin which came out over much the same period but are often undated. Oceanographic apparatus are offered for sale under their headings of meteorological or navigational instruments. Smaller manufacturers' pamphlets survive; sometimes as single or folded sheets describing a single instrument. Much of the apparatus made to ICES standards, constructed by the Scandinavian workshops of Ericsson, Andersen and Olsen, were publicised in this format.

Two British Government commissions were set up to enquire into the technical aspects of laying and operating a submarine cable. The first, which reported in 1861 (Privy Council *Report of the Joint Committee . . .* 1861), dealt with the projected transatlantic cable route via Iceland and Greenland, and the Commission took evidence from several persons well qualified to discuss the practicalities of deep-sea sounding. The second Commission, which enquired into the feasibility of a transpacific route, reported in 1899 (Colonial Office *Report of the Pacific Cable Committee* 1899). By this time the cable companies were undertaking their own surveys and the report highlights the contrast between their efficient methods and the slower and far less accurate surveys that were being carried out by the Hydrographic Department. The annual reports of the Fish Commissions that are recorded in Parliamentary Reports from 1882 onwards give as a rule little detail about the methods of hydrographic research employed. An exception is that of Dickson of 1894 (*12th Annual Report of the Fisheries Board for Scotland*) which gives a vivid picture of scientists trying to carry out their programmes under the joint handicaps of poor instruments and appalling weather.

A number of new journals appeared during the latter half of the nineteenth century; these were devoted to cable engineering, and its related activities including underwater technology; to other aspects of engineering and to general science. Equally important as sources of reference are the governmental and semi-governmental publications from North America and from Britain. I have selected the most useful of these, but items of information often appear in more than one magazine and the numbers being published by the end of the century are far too high to permit listing all possible sources.

1873 *Journal of the Society of Telegraph Engineers and Electricians*
1856 *Engineer*
1866 *Engineering*
1870 *Nature* (London)
1873 *La Nature* (Paris)
1833 *Report of the British Association for the Advancement of Science*
1847 *Report of the US Coast Survey* (later the *US Coast and Geodetic Survey*).

The first great international exhibition of arts and manufactures was staged in London in 1851 and set the pattern for many others, both within Britain and abroad. Products were put on display but there was usually a judging event, and within the publications emanating from any exhibition there are Catalogues, Reports and Awards of Medals to the most worthy exhibitors. Some of these exhibitions were of a very general nature, encompassing every conceivable form of product and manufacture, as well as educational materials and books. Others, however, restricted themselves to one aspect, often related to the commercial life of the city in which they were mounted. Such, for example, are the Fisheries Exhibition of 1880 in Berlin and the Marine and Naval Engineering Exhibition which took place in Glasgow in the same year. This latter event included a series of lectures which were also published. Oceanographic apparatus was of course much in evidence and has been referred to in the text. The Loan Exhibition of Scientific Apparatus, held in London in 1876, displayed a great deal of contemporary and historical material, and British and foreign navies were well represented. The catalogue to this exhibition is very detailed and forms a most useful source of information, regarding the instruments considered at that time to be the most up-to-date and practical of those available. HMS *Challenger* was away on her great scientific voyage, but the Royal Navy took care to include in their collection similar, matching, instruments and gear to that issued to the expedition.

Closely related to the international exhibition, *sensu stricto*, are the Geographical Congresses, which also perambulated from city to city in Europe and America. Here too an exhibition of instruments and cartographic material were often exhibited and may be catalogued at the end of the appropriate congress report. Frederick Ekman showed some of his apparatus in this way and he won an award at the Vienna Congress of 1881.

Technical information concerning the development and use of twentieth century apparatus is to be found in the ICES publications and those of Nansen and his co-workers. ICES put out annually *Rapports et Procès-verbaux* and the irregular *Publications de Circonstance*, as well as other regular reports and statistics relating to hydrography, plankton and fish population studies. The Council's history has been recounted by Arthur

Went (1972) in a special issue of their *Journal*. Many of the summer cruises which Nansen made after the Norwegian North Polar expedition (itself the subject of a series of volumes published in 1901) are described in small books written by Nansen himself or others who went with him. Descriptions of the instruments, the way in which they were used, and the effectiveness of their results are often given, and the narrative is easy to read and interesting.

As marine science became an accepted topic of study, and courses were set up in various universities, a number of formal textbooks appeared. First in the field was Professor J Thoulet of Nancy University, in France, but his two small books were soon joined by the larger volume written by Jules Richard, on the staff of Prince Albert's Oceanographic Museum in Monaco, although his sections dealing with historical apparatus leave much to be desired in their accuracy. The learned and formal volumes of Otto Krümmel were followed by the more popular, though still instructive, works of Herbert Fowler, John Murray and Johan Hjørt. Many other books appeared, devoted to the science alone, but those listed below describe the form and use of instruments and apparatus then in use for research, with which students would be expected to familiarise themselves.

Thoulet J 1895 *Guide d'Océanographie Practique;* 1908 *Instruments et Opérations d'Océanographie Practique*
Richard J 1907 *L'Océanographie*
Krümmel O 1907, 1911 *Handbuch der Ozeanographie*
Fowler H 1912 *Science of the Sea*
Murray J and Hjørt J 1912 *The Depths of the Ocean*

The last source of information on historical oceanographic apparatus brings us right up to the present day. Large collections of instruments survive and are still preserved in various museums around the world. The most numerous is undoubtedly that kept at the Oceanographic Museum in Monaco. Founded by Prince Albert of Monaco in 1902, the Museum opened its doors in 1910, and holds not only the majority of apparatus which the oceanographer Prince used himself but also a fine collection which he was given by foreign workers. It presents therefore examples of virtually every instrument used during the last quarter of the nineteenth and the first years of the twentieth centuries. Replicas were also made of some of the famous instruments of earlier centuries. Chemical apparatus used in ship-board laboratories and biological collecting gear are preserved here but, regrettably, the collection is neither catalogued nor at the time of writing is it well labelled, though plans are afoot to redisplay it, thereby increasing its value to historians and the general visitor.

A second major collection is held at the Science Museum in London. Many of the instruments shown came from the 1876 Loan Exhibition, at which date the Admiralty passed them over to the Museum's keeping. The majority are on display; both these and the remainder in store have been listed in the form of a simple catalogue.

Smaller collections are to be found at the National Maritime Museum in Greenwich, London; at the Royal Scottish Museum in Edinburgh, in the Smithsonian Museum, Washington, at the Geophysical Institute in Bergen, Norway, and at Nansen's house and on his ship *Fram*, both preserved in Oslo, Norway, and open to the public as a memorial to his exploits. Many other maritime and naval museums hold one or two pieces of oceanographic apparatus which have come their way.

Index of Ships

References to page numbers are given in upright type; figure numbers are given in *italics.*

General Index

References to page numbers are given in upright type; figure numbers are given in *italics*.